GED PREP Xcelerator Mathematics

STUDENT BOOK

PAXEN

Melbourne, Florida
www.paxen.com

Acknowledgements

For each of the selections and images listed below, grateful acknowledgement is made for permission to excerpt and/or reprint original or copyrighted material, as follows:

Images

(cover, astronomical clock) Fribus Ekaterina, Shutterstock Images. (cover, architecture) iStockphoto. (cover, drawing compass) Carlos Alvarez, iStockphoto. (cover, circuit board) iStockphoto. v iStockphoto. vi iStockphoto. BLIND Used with the permission of Gil Coronado. 28 Used with the permission of Christopher Blizzard. 56 Used with the permission of Philip Emeagwali. 84 Used with the permission of Huong McDoniel.

ISBN-13: 978-1-934350-24-9
ISBN-10: 1-934350-24-9

2 3 4 5 6 7 8 9 10 GEDXSE2 16 15 14 13 12 11 10   Printed in the U.S.A.

GED PREP XCELERATOR

Mathematics Student Book

Table of Contents

About the GED Tests

Simply by turning to this page, you've made a decision that will change your life for the better. Each year, thousands of people just like you decide to pursue the General Educational Development (GED) certificate. Like you, they left school for one reason or another. And now, just like them, you've decided to continue your education by studying for and taking the GED Tests.

However, the GED Tests are no easy task. The tests—five in all, spread across the subject areas of Language Arts/Reading, Language Arts/Writing, Mathematics, Science, and Social Studies—cover slightly more than seven hours. Preparation takes considerably longer. The payoff, however, is significant: more and better career options, higher earnings, and the sense of achievement that comes with a GED certificate. Employers and colleges and universities accept the GED certificate as they would a high school diploma. On average, GED recipients earn $4,000 more per year than do employees without a GED certificate.

The American Council on Education (ACE) has constructed the GED Tests to mirror a high school curriculum. Although you will not need to know all of the information typically taught in high school, you will need to answer a variety of questions in specific subject areas. In Language Arts/Writing, you will need to write an essay on a topic of general knowledge.

In all cases, you will need to effectively read and follow directions, correctly interpret questions, and critically examine answer options. The table below details the five subject areas, the number of questions within each of them, and the time that you will have to answer them. Since different states have different requirements for the number of tests you may take in a single day, you will need to check with your local adult education center for requirements in your state or territory.

The original GED Tests were released in 1942 and have since been revised a total of three times. In each case, revisions to the tests have occurred as a result of educational findings or workplace needs. All told, more than 17 million people have received a GED certificate since the tests' inception.

SUBJECT AREA TEST	CONTENT AREAS	ITEMS	TIME LIMIT
Language Arts/Reading	Literary texts—75% Nonfiction texts—25%	40 questions	65 minutes
Language Arts/Writing (Editing)	Organization—15% Sentence Structure—30% Usage—30% Mechanics—25%	50 questions	75 minutes
Language Arts/Writing (Essay)	Essay	Essay	45 minutes
Mathematics	Number Sense/Operations—20% to 30% Data Measurement/Analysis—20% to 30% Algebra—20% to 30% Geometry—20% to 30%	Part I: 25 questions (with calculator) Part II: 25 questions (without calculator)	90 minutes
Science	Life Science—45% Earth/Space Science—20% Physical Science—35%	50 questions	80 minutes
Social Studies	Geography—15% U.S. History—25% World History—15% U.S. Government/Civics—25% Economics—20%	50 questions	70 minutes

Three of the subject-area tests—Language Arts/Reading, Science, and Social Studies—will require you to answer questions by interpreting passages. The Science and Social Studies tests also require you to interpret tables, charts, graphs, diagrams, timelines, political cartoons, and other visuals. In Language Arts/Reading, you also will need to answer questions based on workplace and consumer texts. The Mathematics Test will require you to use basic computation, analysis, and reasoning skills to solve a variety of word problems, many of them involving graphics. On all of the tests, questions will be multiple-choice with five answer options. An example follows:

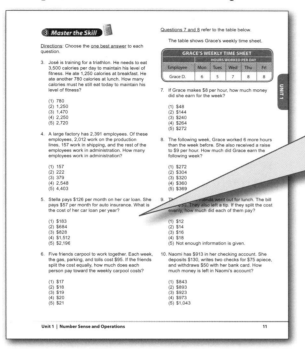

On the Mathematics Test, you will have additional ways in which to register your responses to multiple-choice questions. See p. ix for more information about the three ways of recording responses on the GED Mathematics Test.

As the table on p. iv indicates, the Language Arts/ Writing Test contains two parts, one for editing and the other for essay. In the editing portion of Language Arts/ Writing, you will be asked to identify and correct common errors in various passages and texts while also deciding on the most effective organization of a text. In the essay portion, you will write an essay that provides an explanation or an opinion on a single topic of general knowledge.

So now that you understand the task at hand—and the benefits of a GED certificate—you must prepare for the GED Tests. In the pages that follow, you will find a recipe of sorts that, if followed, will help guide you toward successful completion of your GED certificate. So turn the page. The next chapter of your life begins right now.

About *GED Prep Xcelerator*

Along with choosing to pursue your GED certificate, you've made another smart decision by selecting *GED Prep Xcelerator* as your main study and preparation tool. Simply by purchasing *GED Prep Xcelerator*, you've joined an elite club with thousands of members, all with a common goal—earning their GED certificates. In this case, membership most definitely has its privileges.

For more than 65 years, the GED Tests have offered a second chance to people who need it most. To date, 17 million Americans like you have studied for and earned GED certificates and, in so doing, jump-started their lives and careers. Benefits abound for GED holders: Recent studies have shown that people with GED certificates earn more money, enjoy better health, and exhibit greater interest in and understanding of the world around them than do those without.

In addition, more than 60 percent of GED recipients plan to further their educations, which will provide them with more and better options. As if to underscore the point, U.S. Department of Labor projections show that 90 percent of the fastest growing jobs through 2014 will require postsecondary education.

Your pathway to the future—a *brighter* future—begins now, on this page, with *GED Prep Xcelerator*, an intense, accelerated approach to GED preparation. Unlike other programs, which take months to teach the GED Tests through a content-based approach, *Xcelerator* gets to the heart of the GED Tests—and quickly—by emphasizing *concepts*. At their core, the majority of the GED Tests are reading-comprehension exams. Students must be able to read and interpret excerpts, passages, and various visuals—tables, charts, graphs, timelines, and so on—and then answer questions based upon them.

Xcelerator shows you the way. By emphasizing key reading and thinking concepts, *Xcelerator* equips learners like you with the skills and strategies you'll need to correctly interpret and answer questions on the GED Tests. Two-page micro-lessons in each student book provide focused and efficient instruction, while callout boxes, sample exercises, and test-taking and other thinking strategies aid in understanding complex concepts. For those who require additional support, we offer the *Xcelerator* workbooks, which provide *twice* the support and practice exercises as the student books.

Unlike other GED materials, which were designed *for* the classroom, *Xcelerator* materials were designed *from* the classroom, using proven educational theory and cutting-edge classroom philosophy. The result: More than 90 percent of people who study with *Xcelerator* earn their GED certificates. For learners who have long had the deck stacked against them, the odds are finally in their favor. And yours.

GED BY THE NUMBERS

17 million
Number of GED recipients since the inception of the GED Tests

1.23 million
Number of students who fail to graduate from high school each year

700,000
Number of GED test-takers each year

451,759
Total number of students who passed the GED Tests in 2007

$4,000
Additional earnings per year for GED recipients

About *GED Prep Xcelerator Mathematics*

For those who think the GED Mathematics Test is a breeze, think again. The GED Mathematics Test is a rigorous exam that will assess your ability to answer a range of mathematics questions about various topics, such as estimation, ratio, and proportion. About half of the GED Mathematics Test involves answering questions using tables, charts, graphs, diagrams, and drawings. Tables, for example, organize group complex data using rows and columns. As you can see on the sample page below, rows are read from top to bottom, and columns from left to right. Tables often include headings that describe the data they contain.

All told, you will have a total of 90 minutes in which to answer 50 multiple-choice items organized across four main content areas, each of which make up between 20% and 30% of all questions: Number Sense and Operations; Measurement/Data Analysis; Algebra, Patterns, and Functions; and Geometry. Material in *GED Prep Xcelerator Mathematics* has been organized with these percentages in mind.

GED Prep Xcelerator Mathematics helps learners like you build and develop core mathematics skills. A combination of targeted strategies, informational callouts and sample questions, assorted test-taking tips, and ample assessment help to clearly focus study efforts in needed areas, all with an eye toward success on the GED Tests.

You will use a site-issued calculator to answer questions in Part I of the GED Mathematics Test. In addition, a formulas page, such as that on p. viii in this book, will be supplied. It includes all of the formulas needed to succeed on the GED Mathematics Test.

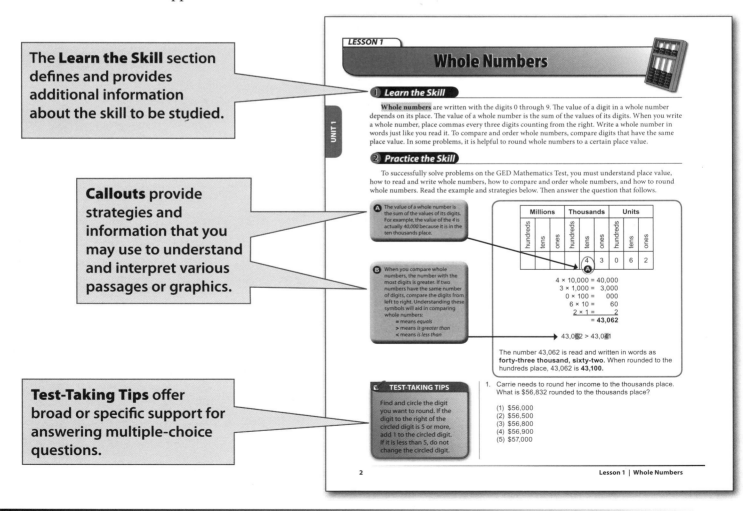

The **Learn the Skill** section defines and provides additional information about the skill to be studied.

Callouts provide strategies and information that you may use to understand and interpret various passages or graphics.

Test-Taking Tips offer broad or specific support for answering multiple-choice questions.

Formulas for GED Mathematics Test

Area of a...

Square:	Area = side²
Rectangle:	Area = length × width
Parallelogram:	Area = base × height
Triangle:	Area = $\frac{1}{2}$ × base × height
Trapezoid:	Area = $\frac{1}{2}$ × (base$_1$ + base$_2$) × height
Circle:	Area = π × radius²; π equals about 3.14

Perimeter of a...

Square:	Perimeter = 4 × side
Rectangle:	Perimeter = 2 × length + 2 × width
Triangle:	Perimeter = side$_1$ + side$_2$ + side$_3$

Circumference of a...

Circle:	Circumference = π × diameter

Volume of a...

Cube:	Volume = edge³
Rectangular prism:	Volume = length × width × height
Square pyramid:	Volume = $\frac{1}{3}$ × (base edge)² × height
Cylinder:	Volume = π × radius² × height
Cone:	Volume = $\frac{1}{3}$ × π × radius² × height

Coordinate geometry

Distance between points = $\sqrt{(x_2 - x_1)^2 + (y_2 - y_1)^2}$

Slope of a line = $\frac{y_2 - y_1}{x_2 - x_1}$

Pythagorean relationship

$a^2 + b^2 = c^2$; a and b are legs, and c is the hypotenuse

Central measures

Mean = $\frac{x1 + x2 + ... + xn}{n}$;		where x's are the individual values, and n is the total number of values for x.
Median =		The middle value of an odd number of ordered scores, and halfway between the two middle values of an even number of ordered scores.
Mode =		The most common number in a set.

Simple interest	**Interest** =	Principal × rate × time
Distance	**Distance** =	Rate × time
Total cost	**Total cost** =	(Number of units) × (Price per unit)

Response Options for the GED Mathematics Test

Once you read each question, apply any formulas, and determine the correct answer, you then must mark the answer correctly on the GED answer sheet. On the GED Mathematics Test, you will have three options to use, all of which will be explained below and on the test itself. The Unit Reviews in *GED Prep Xcelerator Mathematics* have been organized to provide you with practice in using all three response styles.

The horizontal-response format requires you to fill in the numbered circle that corresponds to the answer you select for each question on the GED Mathematics Test. Mark your answer completely, making no stray or unnecessary marks. If you change an answer, erase your first mark completely. Mark only one answer space for each question; multiple answers will be scored as incorrect.

To record your answers in the alternate-response format, you must do the following:
- Begin in any column that will allow your answer to be entered;
- Write your answer in the boxes in the top row;
- In the column beneath a fraction bar or decimal point (if any) and each number in your answer, fill in the bubble representing that character;
- Leave blank any unused columns.

Points to consider when recording an answer on the coordinate grid:
- To record an answer, you must have an *x* and *y* value.
- No answer will have a value that is a fraction or a decimal.
- Mark only one circle that represents your answer.

Calculator Directions

The GED Mathematics Test has two sections, one of which allows for the use of a calculator. Therefore, calculators are permitted in certain lessons of this book. When you see a 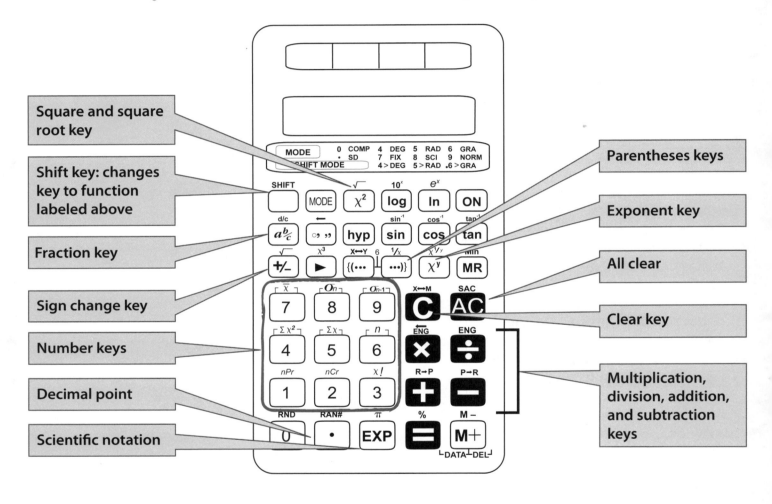 symbol within a *GED Prep Xcelerator* student book or workbook lesson, this means that calculator use is allowed. When you do not see a symbol, then calculators should not be used. The Unit Reviews contain a representative sampling of problems from all lessons in a unit. Therefore, each item in a Unit Review that permits the use of a calculator will show a symbol.

The Casio FX-260 Solar is the featured calculator for the GED Mathematics Test. Your testing center will provide this calculator when you take the test. The calculator is shown below, along with callouts that indicate some of its most important keys.

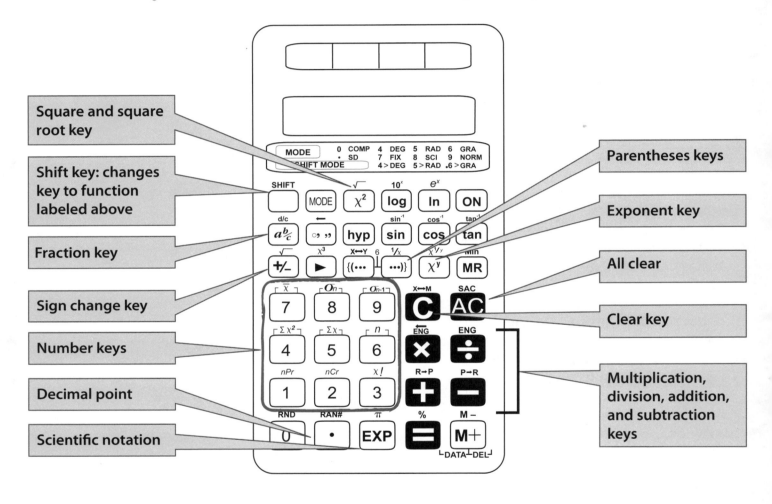

Square and square root key

Shift key: changes key to function labeled above

Fraction key

Sign change key

Number keys

Decimal point

Scientific notation

Parentheses keys

Exponent key

All clear

Clear key

Multiplication, division, addition, and subtraction keys

Getting Started

Press the **ON** key to turn on the calculator. Check the display screen for "DEG" in the upper center and 0 at the right. If you do not see "DEG," then press **MODE** 4.

- Use the **AC** key when you wish to clear all numbers and operations from the screen.
- Use the **C** key when you only wish to delete the last number or operation shown on the screen.

Order of Operations

Note that the calculator uses the order of operations to perform calculations. For example, if you wish to find the mean of 5, 9, 3, 4, and 2, you would need to make sure that you divide the *sum* by 5. If you entered 5 **+** 9 **+** 3 **+** 4 **+** 2 ÷ 5, the calculator would interpret this as $5 + 9 + 3 + 4 + \frac{2}{5}$.

Signed Numbers

Press the **+/−** key when you wish to change a positive number to a negative number, or vice versa. For example, to solve 4 − (−5), enter "4 **−** 5 **+/−** **=**." Note that although the minus symbol and negative sign may look similar on paper, there is a different calculator key for each symbol.

> Use your calculator to simplify −7 − (−9).
> The correct answer is 2.

Parentheses

Use the **((···** and **···))** keys to group numbers and symbols. For example, to solve $\frac{-4 + 9}{5 \times 2}$, enter "**((···** 4 **+/−** **+** 9 **···))** **÷** **((···** 5 **×** 2 **···))** **=**."

> Use your calculator to simplify $9 + \frac{4 - 20}{-3 + 5}$.
> The correct answer is 1.

Squares, Square Roots, and Exponents

Use the **x²** key to raise a number to the second power. For example, to solve $(-6)^2$, enter "6 **+/−** **x²**."

- To find the square root of a number, use the **SHIFT** and **x²** keys. For example, to find the square root of 200, enter "200 **SHIFT** **x²**." The **SHIFT** key changes the function from x^2 to $\sqrt{\ }$.
- To raise a number to a power, use the **xʸ** key. For example, to find 6^4, enter "6 **xʸ** 4."

Note that when squaring numbers or finding square roots, you do not need to press **=**.

> Use your calculator to simplify $3^4 - \sqrt{25} + (-3)^2$.
> The correct answer is 85.

Test-Taking Tips

The GED Tests include 240 questions across the five subject-area exams of Language Arts/Reading, Language Arts/Writing, Mathematics, Science, and Social Studies. In each of the GED Tests, you will need to apply some amount of subject-area knowledge. However, because all of the questions are multiple-choice items largely based on text or visuals (such as tables, charts, or graphs), the emphasis in *GED Prep Xcelerator* is on helping learners like you build and develop core reading and thinking skills. As part of the overall strategy, various test-taking tips are included below and throughout the book to help you improve your performance on the GED Tests. For example:

◆ *Always thoroughly read the directions so that you know exactly what to do.* In Mathematics, for example, one part of the test allows for the use of a calculator. The other part does not. If you are unsure of what to do, ask the test provider if the directions can be explained.

◆ *Read each question carefully so that you fully understand what it is asking.* Some questions, for example, may present more information than you need to correctly answer them. Other questions may note emphasis through capitalized and boldfaced words (Which of the following is **NOT** an example of photosynthesis?).

◆ *Manage your time with each question.* Because the GED Tests are timed exams, you'll want to spend enough time with each question, but not *too* much time. For example, on the GED Mathematics Test, you have 90 minutes in which to answer 50 multiple-choice questions. That works out to less than two minutes per item. You can save time by first reading each question and its answer options before reading the passage or examining the graphic. Once you understand what the question is asking, review the passage or visual for the appropriate information.

◆ *Note any unfamiliar words in questions.* First, attempt to reread the question by omitting the unfamiliar word(s). Next, try to substitute another word in its place.

◆ *Answer all questions, regardless of whether you know the answer or are guessing at it.* There is no benefit in leaving questions unanswered on the GED Tests. Keep in mind the time that you have for each test and manage it accordingly. For time purposes, you may decide to initially skip questions. However, note them with a <u>light mark</u> beside the question and try to return to them before the end of the test.

◆ *Narrow answer options by rereading each question and the text or graphic that goes with it.* Although all five answers are *possible,* keep in mind that only one of them is *correct.* You may be able to eliminate one or two answers immediately; others may take more time and involve the use of either logic or assumptions. In some cases, you may need to make your best guess between two options. If so, keep in mind that test makers often avoid answer patterns; that is, if you know the previous answer is (2) and are unsure of the answer to the next question but have narrowed it to options (2) and (4), you may want to choose (4).

◆ *Read all answer choices.* Even though the first or second answer choice may appear to be correct, be sure to thoroughly read all five answer choices. Then go with your instinct when answering questions. For example, if your first instinct is to mark (1) in response to a question, it's best to stick with that answer unless you later determine that answer to be incorrect. Usually, the first answer you choose is the correct one.

◆ *Correctly complete your answer sheet by marking one numbered space on the answer sheet beside the number that corresponds to it.* Mark only one answer for each item; multiple answers will be scored as incorrect. If time permits, double-check your answer sheet after completing the test to ensure that you have made as many marks—no more, no less—as there are questions.

You've already made two very smart decisions in trying to earn your GED certificate and in purchasing *GED Prep Xcelerator* to help you to do so. The following are additional strategies to help you optimize your success on the GED Tests.

3 weeks out ...

◆ Set a study schedule for the GED Tests. Choose times in which you are most alert, and places, such as a library, that provide the best study environment.

◆ Thoroughly review all material in *GED Prep Xcelerator,* using the *GED Prep Xcelerator Mathematics Workbook* to extend understanding of concepts in the *GED Prep Xcelerator Mathematics Student Book.*

◆ Make sure that you have the necessary tools for the job: sharpened pencils, pens, paper, and, for Mathematics, the Casio FX-260 Solar calculator.

◆ Keep notebooks for each of the subject areas that you are studying. Folders with pockets are useful for storing loose papers.

◆ When taking notes, restate thoughts or ideas in your own words rather than copying them directly from a book. You can phrase these notes as complete sentences, as questions (with answers), or as fragments, provided you understand them.

1 week out ...

◆ Take the pretests, noting any troublesome subject areas. Focus your remaining study around those subject areas.

◆ Prepare the items you will need for the GED Tests: admission ticket (if necessary), acceptable form of identification, some sharpened No. 2 pencils (with erasers), a watch, eyeglasses (if necessary), a sweater or jacket, and a high-protein snack to eat during breaks.

◆ Map out the course to the test center, and visit it a day or two before your scheduled exam. If you drive, find a place to park at the center.

◆ Get a good night's sleep the night before the GED Tests. Studies have shown that learners with sufficient rest perform better in testing situations.

The day of ...

◆ Eat a hearty breakfast high in protein. As with the rest of your body, your brain needs ample energy to perform well.

◆ Arrive 30 minutes early to the testing center. This will allow sufficient time in the event of a change to a different testing classroom.

◆ Pack a sizeable lunch, especially if you plan to be at the testing center most of the day.

◆ Focus and relax. You've come this far, spending weeks preparing and studying for the GED Tests. It's your time to shine.

Unit 1

GIL CORONADO

After earning his GED certificate, Gil Coronado experienced continued success in the military and, later, as head of the Selective Service System.

Gil Coronado has a knack for numbers. Coronado, who left high school to enlist in the United States Air Force, earned his GED certificate and went on to a distinguished 30-year career in the military. During that time, he earned more than 35 awards, including the prestigious Bronze Star for his service during the Vietnam War. As he noted,

> **❝ The military opened more doors in my life than I ever thought existed. It's truly one of the best experiences a young person can have ... ❞**

After retiring from the military, Coronado continued to serve his country as deputy assistant secretary at the Department of Veterans Affairs. In 1994, Coronado accepted a position requiring a knowledge of mathematics and a talent for organization when he was appointed the ninth director of the Selective Service System. Each year, federal law requires all men, about 1.8 million per year, to register with Selective Service within 30 days of turning 18.

As director of the Selective Service, Coronado managed 180 federal employees, 11,000 board members, more than 50 appointed state heads, and 450 reserve officers. He also provided leadership to regional directors and the Data Measurement Center. As the agency's first Hispanic director, Coronado updated and streamlined the system so that men can register through the Internet.

Throughout his career, Coronado remained true to his roots. He advocated for creation of Hispanic Heritage Month, established by Congress in 1988. He founded and chaired the group "Heroes and Heritage" and served as a member of the National Consortium for Education Access. Today, he serves as vice president of Kondor Films, which produces both documentary films and an animated educational series.

BIO BLAST: Gil Coronado

- Born and raised in San Antonio, Texas
- Earned his college degree from Our Lady of the Lake University
- Graduated from several military service schools
- Served in Southeast Asia during the Vietnam War
- Named European Commander of the Year
- Inducted into the U.S. Army Officer Candidate School Hall of Fame

Number Sense and Operations

Unit 1: Number Sense and Operations

You are surrounded by numbers. Whether paying bills, negotiating a car loan, budgeting for rent or groceries, depositing a check, or withdrawing money, you use basic math skills such as addition, subtraction, multiplication, and division to perform a variety of everyday tasks.

In the same way, number sense and operations play an important part on the GED Mathematics Test, making up between 20 and 30 percent of questions. In Unit 1, you will study whole numbers, word problems, fractions, ratios, decimals, and percents, all of which will help you prepare for the GED Mathematics Test.

Table of Contents

Whole Numbers

① Learn the Skill

Whole numbers are written with the digits 0 through 9. The value of a digit in a whole number depends on its place. The value of a whole number is the sum of the values of its digits. When you write a whole number, place commas every three digits counting from the right. Write a whole number in words just like you read it. To compare and order whole numbers, compare digits that have the same place value. In some problems, it is helpful to round whole numbers to a certain place value.

② Practice the Skill

To successfully solve problems on the GED Mathematics Test, you must understand place value, how to read and write whole numbers, how to compare and order whole numbers, and how to round whole numbers. Read the example and strategies below. Then answer the question that follows.

Ⓐ The value of a whole number is the sum of the values of its digits. For example, the value of the *4* is actually *40,000* because it is in the ten thousands place.

Ⓑ When you compare whole numbers, the number with the most digits is greater. If two numbers have the same number of digits, compare the digits from left to right. Understanding these symbols will aid in comparing whole numbers:

> = means *equals*
> > means *is greater than*
> < means *is less than*

Millions			Thousands			Units		
hundreds	tens	ones	hundreds	tens	ones	hundreds	tens	ones
				④	3	0	6	2

Ⓐ

$$4 \times 10{,}000 = 40{,}000$$
$$3 \times 1{,}000 = 3{,}000$$
$$0 \times 100 = 000$$
$$6 \times 10 = 60$$
$$\underline{2 \times 1 = 2}$$
$$= \mathbf{43{,}062}$$

43,0**6**2 > 43,0**4**1

The number 43,062 is read and written in words as **forty-three thousand, sixty-two.** When rounded to the hundreds place, 43,062 is **43,100.**

☑ TEST-TAKING TIPS

Find and circle the digit you want to round. If the digit to the right of the circled digit is 5 or more, add 1 to the circled digit. If it is less than 5, do not change the circled digit.

1. Carrie needs to round her income to the thousands place. What is $56,832 rounded to the thousands place?

(1) $56,000
(2) $56,500
(3) $56,800
(4) $56,900
(5) $57,000

UNIT 1

Directions: Choose the one best answer to each question.

2. Meredith wrote a check for $182 to pay a bill. How is 182 written in words?

    (1) one hundred eight-two
    (2) one hundred eighty-two
    (3) one hundred and eighteen-two
    (4) one-hundred eighty and two
    (5) one-hundred eight tens and two

3. Mr. Murphy rounds his students' test scores to the tens place. Jonathan's test score is 86. What is his test score rounded to the tens place?

    (1) 80
    (2) 85
    (3) 86
    (4) 90
    (5) 100

4. Each book in a historical library is given a number. The books are arranged on shelves according to their numbers. The range of numbers for shelves H through L is shown below.

    Shelf H 1245–1336
    Shelf I   1337–1420
    Shelf J 1421–1499
    Shelf K 1500–1622
    Shelf L 1623–1708

    On which shelf would you find the book that is numbered 1384?

    (1) Shelf H
    (2) Shelf I
    (3) Shelf J
    (4) Shelf K
    (5) Shelf L

5. Michael swam 2,450 yards on Monday, 2,700 yards on Tuesday, and 2,250 yards on Wednesday. What is the order of his daily swim yardage from least to greatest?

    (1) 2,450, 2,700, 2,250
    (2) 2,250, 2,700, 2,450
    (3) 2,250, 2,450, 2,700
    (4) 2,700, 2,450, 2,250
    (5) 2,450, 2,250, 2,700

Questions 6 and 7 refer to the information and table below.

The table below shows a sporting goods store's monthly sales for the first six months of the year.

MONTHLY SALES	
January	$155,987
February	$150,403
March	$139,605
April	$144,299
May	$149,355
June	$148,260

6. Based on the table, in which month did the store have its highest sales?

    (1) January
    (2) February
    (3) March
    (4) April
    (5) May

7. Based on the table, what sales trend can you determine?

    (1) People purchased the most sporting goods equipment during early spring.
    (2) Sales were at their highest in winter months.
    (3) Monthly sales remained the same from January through June.
    (4) People purchased more sporting goods as summer approached.
    (5) Sales suffered in the months immediately following the winter holidays.

Operations

① Learn the Skill

The four basic math operations are addition, subtraction, multiplication, and division. Add quantities to find a total, or **sum**. Use subtraction to find the **difference** between two quantities.

Multiply to find a **product** when you need to add a number many times. Divide when separating a quantity into equal groups. The initial quantity is the **dividend**. The **divisor** is the number by which you divide. The **quotient** is the answer.

② Practice the Skill

To successfully solve problems on the GED Mathematics Test, you must determine the correct operation(s) to perform and the proper order in which to perform them. Read the examples and strategies below. Then answer the question that follows.

Ⓐ Add the numbers in each column, working from right to left. If the sum of a column of digits is greater than 9, regroup to the next column on the left.

Ⓑ To subtract, align the digits by place value. Subtract the numbers in each column, working from right to left. When a digit in the bottom number is greater than the digit in the top number, regroup.

Ⓒ Multiply the ones digit of the bottom number by all the digits in the top number. Align each result, or partial product, under the digit you multiplied by. Use zeros as placeholders. When you have multiplied all digits in the bottom number by all digits in the top number, add the partial products.

Ⓐ Addition

$$\begin{array}{r} \overset{1}{4}82 \\ + 208 \\ \hline \mathbf{690} \end{array}$$

Ⓒ Multiplication

$$\begin{array}{r} 482 \\ \times\ \ 34 \\ \hline 1,928 \\ \times\ 14,460 \\ \hline \mathbf{16,388} \end{array}$$

Ⓑ Subtraction

$$\begin{array}{r} 4\overset{7}{\cancel{8}}\overset{12}{\cancel{2}} \\ - 208 \\ \hline \mathbf{274} \end{array}$$

Ⓓ Division

$$\begin{array}{r} \mathbf{517\ R12} \\ 14\overline{)7250} \\ -\underline{70} \\ 25 \\ -\underline{14} \\ 110 \\ -\underline{98} \\ 12 \end{array}$$

Ⓓ If the first digit of the dividend is greater than or equal to the divisor, divide. If not, divide the first two (or if needed, three) digits of the dividend by the divisor. Decide how many divisor groups are contained in the partial dividend. Write that number in the quotient, being sure to align place values. Multiply this number by the divisor and write the product under the part of the dividend that you divided. Repeat until all digits in the dividend have been used. Write any remainder next to the quotient.

☑ TEST-TAKING TIPS

To determine which operation to use to solve a problem, look for key words or phrases. Words like *total* and *sum* indicate addition. Phrases like *how much is left, less than,* and *difference* indicate subtraction.

1. Shirley has $1,256 in her bank account. She withdraws $340. How much money is left in her bank account?

   (1) $816
   (2) $916
   (3) $926
   (4) $996
   (5) $1,006

Directions: Choose the <u>one best answer</u> to each question.

2. Alex drove from Denver, Colorado, to Chicago, Illinois, in two days. The first day he drove 467 miles. The second day he drove 583 miles. What is the total distance that Alex drove?

   (1) 950 miles
   (2) 1,039 miles
   (3) 1,040 miles
   (4) 1,049 miles
   (5) 1,050 miles

3. During a word game, Alicia had 307 points. She was unable to use all of her letters, so she had to subtract 19 points at the end of the game. What was Alicia's final score?

   (1) 287
   (2) 288
   (3) 297
   (4) 298
   (5) 307

4. Juan works 40 hours per week. He earns $9 per hour. How much does Juan earn in one week?

   (1) $32
   (2) $36
   (3) $320
   (4) $360
   (5) $400

5. Carl pays $45 per month for car insurance. How much does he spend on car insurance in 1 year?

   (1) $550
   (2) $540
   (3) $530
   (4) $450
   (5) $440

6. Four friends went out for pizza. The total cost for appetizers, pizza, and drinks was $64. If the friends split the cost equally, how much did each friend pay?

   (1) $13
   (2) $14
   (3) $15
   (4) $16
   (5) $17

Question 7 refers to the following diagram.

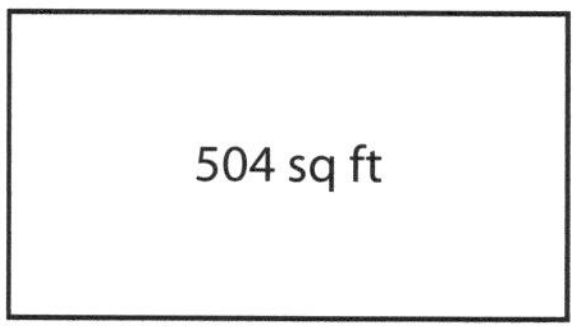

504 sq ft

7. Claire is purchasing bags of mulch to cover her vegetable garden. One bag of mulch will cover 12 square feet. How many bags of mulch will Claire need?

   (1) 41
   (2) 42
   (3) 43
   (4) 44
   (5) 45

8. Each month, Anna pays $630 in rent. How much rent does she pay over the course of 18 months?

   (1) $1,340
   (2) $6,300
   (3) $11,340
   (4) $12,600
   (5) $15,120

9. The quarterback on Scott's favorite football team is closing in on a 4,000-yard passing season. He has thrown for 3,518 yards with two games remaining. How many yards would the quarterback need to average over the final two games to reach his goal of 4,000 yards?

   (1) 221
   (2) 241
   (3) 271
   (4) 311
   (5) 482

Word Problems

1 Learn the Skill

There are several steps to solving word problems. First, read the problem carefully. Make sure you understand all of the information. Next, determine the information that you need to solve the problem. There may be more information than you need. In such cases, you must decide which information is necessary to solve the problem.

2 Practice the Skill

Since many of the questions on the GED Mathematics Test are word problems, understanding how to quickly and accurately solve such problems will improve your ability to complete the test successfully. Read the example and strategies below. Then answer the question that follows.

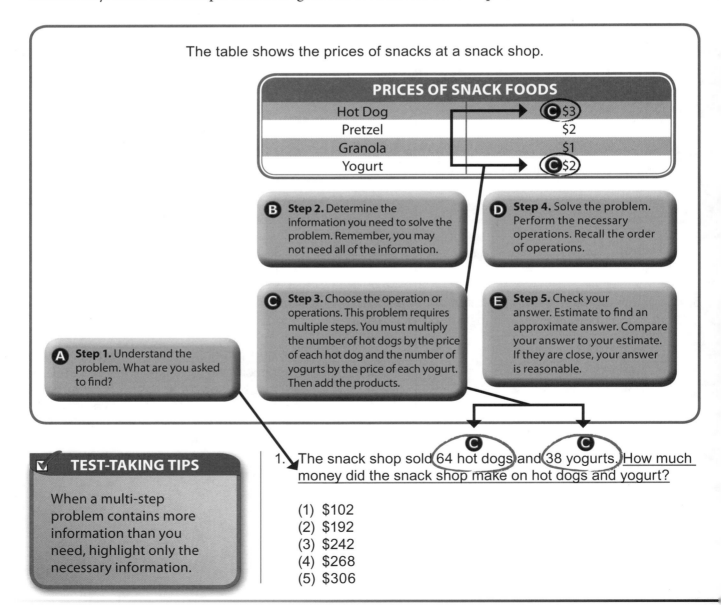

The table shows the prices of snacks at a snack shop.

PRICES OF SNACK FOODS

Hot Dog	**C** $3
Pretzel	$2
Granola	$1
Yogurt	**C** $2

B Step 2. Determine the information you need to solve the problem. Remember, you may not need all of the information.

D Step 4. Solve the problem. Perform the necessary operations. Recall the order of operations.

C Step 3. Choose the operation or operations. This problem requires multiple steps. You must multiply the number of hot dogs by the price of each hot dog and the number of yogurts by the price of each yogurt. Then add the products.

E Step 5. Check your answer. Estimate to find an approximate answer. Compare your answer to your estimate. If they are close, your answer is reasonable.

A Step 1. Understand the problem. What are you asked to find?

TEST-TAKING TIPS

When a multi-step problem contains more information than you need, highlight only the necessary information.

1. The snack shop sold 64 hot dogs and 38 yogurts. How much money did the snack shop make on hot dogs and yogurt?

(1) $102
(2) $192
(3) $242
(4) $268
(5) $306

Directions: Choose the one best answer to each question.

2. Allison wants to buy a pair of shoes that originally cost $54. Today the shoes are on sale for $42. She uses a coupon for $5 off a pair of shoes. How much does Allison pay for the shoes?

   (1) $12
   (2) $37
   (3) $42
   (4) $47
   (5) $54

3. Rasheed pays $645 each month for rent. He pays $78 each month for heat and electricity and $25 each month for phone service. What is the total amount he pays each month for rent, heat and electricity, and phone service?

   (1) $567
   (2) $620
   (3) $670
   (4) $723
   (5) $748

4. In the fall, 143 children signed up to play after-school sports at an elementary school. Sixty-seven children signed up for soccer, and thirty-three children signed up for softball. How many children signed up for a sport other than soccer?

   (1) 76
   (2) 110
   (3) 176
   (4) 210
   (5) Not enough information is given.

Questions 5 and 6 refer to the following table.

The following table shows the cost that a store pays for clothing and the price at which the store sells the clothing.

PRICES OF CLOTHING		
ITEM	AMOUNT STORE PAYS	PRICE STORE CHARGES
Women's T-shirt	$6	$12
Men's T-shirt	$6	$10
Women's sport shorts	$7	$14
Men's sport shorts	$9	$16

5. The store paid a bill to its supplier for a shipment of 144 women's T-shirts. What was the amount of the bill?

   (1) $720
   (2) $844
   (3) $864
   (4) $1,008
   (5) $1,728

6. The profit the store makes on each item of clothing is the difference between what the store pays for the item and the price at which the store sells the item. If the store sells 508 men's sport shorts, what is the store's profit?

   (1) $3,556
   (2) $4,064
   (3) $4,572
   (4) $7,112
   (5) $8,128

7. The Stein family budgeted $5,500 for their vacation to cover airfare, lodging, meals, and extras, such as souvenirs. Their total for airfare, lodging, and meals came to $5,000. Of the remainder, what was the average amount that each of the four family members could spend on souvenirs and remain within their budget?

   (1) $50
   (2) $125
   (3) $150
   (4) $250
   (5) $500

UNIT 1

Fractions

① Learn the Skill

A **fraction** shows part of a whole or part of a group by separating two numbers with a fraction bar. The bottom number is called the **denominator**. It tells the number of equal parts in a whole. The top number is the **numerator**. It tells the number of equal parts being considered.

② Practice the Skill

You will use and compare proper fractions, improper fractions, and mixed numbers on the GED Mathematics Test. Always write a fraction in lowest terms. For example, in the fraction $\frac{8}{10}$, both the numerator and denominator can be divided by the same number, 2. $\frac{8}{10}$ written in lowest terms is $\frac{4}{5}$. Read the examples and strategies below. Then answer the question that follows.

A A proper fraction shows a quantity less than 1, such as $\frac{4}{5}$. A mixed number is made up of a whole number and a proper fraction. An improper fraction can be written as a mixed number by dividing the numerator by the denominator. The whole number is the whole-number part of the mixed number. Write the remainder as the numerator over the original denominator. To write a mixed number as an improper fraction, multiply the denominator by the whole number and add the numerator. Write this number over the original denominator.

Proper Fractions
Determine the shaded portion of the circle graph in fraction form.

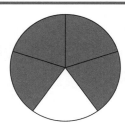

$\dfrac{4}{5}$ ←numerator
←denominator

Order $\frac{2}{5}$, $\frac{3}{4}$, $\frac{4}{5}$, and $\frac{1}{4}$ from least to greatest.

$$\frac{2 \times 4}{5 \times 4} = \frac{8}{20} \quad \Big| \quad \frac{3 \times 5}{4 \times 5} = \frac{15}{20} \quad \Big| \quad \frac{4 \times 4}{5 \times 4} = \frac{16}{20} \quad \Big| \quad \frac{1 \times 5}{4 \times 5} = \frac{5}{20}$$

$$\frac{1}{4} < \frac{2}{5} < \frac{3}{4} < \frac{4}{5}$$

Mixed Numbers

Write $\frac{7}{5}$ as a mixed number.

$$5\overline{)7} \quad 1\frac{2}{5}$$
$$\underline{-5}$$
$$2$$

Improper Fractions

Write $1\frac{2}{5}$ as an improper fraction.

$$1\frac{2}{5} = \frac{(5 \times 1) + 2}{5} = \frac{7}{5}$$

☑ **TEST-TAKING TIPS**

If you struggle to find the lowest common denominator, try multiplying the denominators by each other.

1. Aliah ran $\frac{2}{5}$ of a mile. Charlie ran $\frac{3}{4}$ of a mile. Ella ran $\frac{4}{5}$ of a mile, and Carlos ran $\frac{1}{4}$ of a mile. Who ran the farthest?

(1) Aliah
(2) Charlie
(3) Ella
(4) Carlos
(5) Not enough information is given.

Directions: Choose the <u>one best answer</u> to each question.

<u>Questions 2 and 3</u> refer to the following table.

In a water relay race, each team must fill a cup of water, race over to a bowl, and pour the water from the cup into the bowl. The relay is over when one team has filled its bowl to the top. The table below shows the results of the race.

WATER RELAY RESULTS	
Team	**Bowl Capacity**
Team 1	$\frac{1}{2}$
Team 2	$\frac{1}{1}$
Team 3	$\frac{3}{5}$
Team 4	$\frac{1}{3}$
Team 5	$\frac{4}{5}$

2. Which team had a bowl less than $\frac{1}{2}$ full?

   (1) Team 1
   (2) Team 2
   (3) Team 3
   (4) Team 4
   (5) Team 5

3. Which team won the relay race?

   (1) Team 1
   (2) Team 2
   (3) Team 3
   (4) Team 4
   (5) Team 5

4. Jenny needs to add $2\frac{3}{4}$ and $1\frac{5}{8}$. She must find a common denominator. What is the lowest common denominator of 4 and 8?

   (1) 2
   (2) 4
   (3) 8
   (4) 12
   (5) 16

5. Clark is baking cookies. He needs $2\frac{1}{2}$ cups of flour. How many times will he have to fill his $\frac{1}{2}$-cup measuring cup to equal $2\frac{1}{2}$ cups?

   (1) 1
   (2) 2
   (3) 3
   (4) 4
   (5) 5

6. Elyse is running for a seat on the local school board. She received $\frac{1}{4}$ of the 200 votes cast. Another contender, James, received $\frac{2}{5}$ of the vote. A third candidate, Michele, received $\frac{3}{10}$ of the vote. A final candidate, Tom, received $\frac{1}{20}$ of the vote. Who won the election?

   (1) Elyse
   (2) James
   (3) Michele
   (4) Tom
   (5) Not enough information is given.

<u>Question 7</u> refers to the following table.

Five friends in a study group ordered pizza. The number of slices they ate is listed below.

AMOUNT OF PIZZA	
Person	**Number of Slices**
Naomi	$\frac{5}{2}$
Marty	$\frac{7}{4}$
Sam	$\frac{4}{3}$
Abigail	$1\frac{5}{8}$
Kirsten	$2\frac{1}{3}$

7. Which friend ate the most pizza?

   (1) Naomi
   (2) Marty
   (3) Sam
   (4) Abigail
   (5) Kirsten

Operations with Fractions

① Learn the Skill

Fractions and mixed numbers can be added, subtracted, multiplied, and divided. To add or subtract like fractions, simply add or subtract the numerators and keep the denominator. To add or subtract unlike fractions, first find a common denominator. Common denominators are not needed for multiplication or division.

② Practice the Skill

You must understand how to perform operations with fractions and mixed numbers to correctly solve problems on the GED Mathematics Test. You also will need to determine the proper operation to use to solve fraction problems. Read the examples and strategies below. Then answer the question that follows.

A To multiply fractions, simply multiply the numerators and multiply the denominators. Reduce the product to lowest terms. To divide, multiply by the reciprocal of the divisor. Write whole numbers as fractions by writing them over 1. For example, $9 = \frac{9}{1}$

B To add mixed numbers, first find a common denominator. Then add the fractions. If the sum is an improper fraction, change it to a mixed number. Then add the whole numbers and the sum of the fractions to the sum of the whole numbers.

Add $\quad \frac{3}{4} + \frac{5}{8} \rightarrow \frac{3 \times 2}{4 \times 2} = \frac{6}{8} \quad \frac{6}{8} + \frac{5}{8} = \frac{11}{8} = 1\frac{3}{8}$

Subtract $\frac{3}{4} - \frac{5}{8} \rightarrow \frac{3 \times 2}{4 \times 2} = \frac{6}{8} \quad \frac{6}{8} - \frac{5}{8} = \frac{1}{8}$

Multiply $\frac{3}{4} \times \frac{5}{8} \rightarrow \frac{3}{4} \times \frac{5}{8} = \frac{15}{32}$

Divide $\quad \frac{5}{9} \div \frac{2}{3} \rightarrow \frac{5}{9} \div \frac{2}{3} = \frac{5}{9} \times \frac{3}{2} = \frac{15}{18} = \frac{5}{6}$

Add $\quad 4\frac{5}{6} + 2\frac{1}{4}$

$4\frac{5}{6} + 2\frac{1}{4} = 4\frac{5 \times 2}{6 \times 2} + 2\frac{1 \times 3}{4 \times 3} = 4\frac{10}{12} + 2\frac{3}{12} = 6\frac{13}{12} = 7\frac{1}{12}$

☑ TEST-TAKING TIPS

To multiply and divide mixed numbers, first rename the mixed numbers as improper fractions.

1. There are two containers of milk in Eric's refrigerator. One has $\frac{3}{5}$ gallon of milk. The other has $\frac{3}{4}$ gallon of milk. How many gallons of milk are in Eric's refrigerator?

   (1) $\frac{9}{20}$

   (2) $\frac{6}{11}$

   (3) $1\frac{7}{20}$

   (4) $1\frac{9}{20}$

   (5) $1\frac{3}{5}$

UNIT 1

③ **Apply the Skill**

Directions: Choose the <u>one best answer</u> to each question.

2. Carly spent $3\frac{1}{3}$ hours organizing her room. She spent $1\frac{1}{2}$ hours cleaning her room. How many hours did she spend on her room altogether?

   (1) $1\frac{5}{6}$

   (2) $2\frac{1}{3}$

   (3) $4\frac{1}{6}$

   (4) $4\frac{2}{3}$

   (5) $4\frac{5}{6}$

3. Blake needs a wooden dowel that is $2\frac{1}{6}$ feet long. How much should he cut off the dowel shown below?

   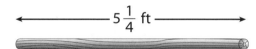
   $5\frac{1}{4}$ ft

   (1) $3\frac{1}{4}$ feet

   (2) $3\frac{1}{6}$ feet

   (3) $3\frac{1}{12}$ feet

   (4) $2\frac{1}{10}$ feet

   (5) $2\frac{1}{12}$ feet

4. A recipe calls for $1\frac{3}{4}$ cups of flour. If Liza cuts the recipe in half, how much flour will she use?

   (1) $\frac{7}{8}$ cup

   (2) 1 cup

   (3) $1\frac{1}{4}$ cups

   (4) $2\frac{1}{4}$ cups

   (5) $3\frac{1}{2}$ cups

5. Mr. White is sewing new curtains. He has a piece of material that is $11\frac{1}{4}$ yards long. He needs $2\frac{1}{4}$ yards for each curtain. How many curtains can he make from this material?

   (1) 5

   (2) 4

   (3) 3

   (4) 2

   (5) 1

6. Colin has a stack of 8 books on his desk. If each book is $\frac{2}{3}$ of an inch thick, what is the height of the stack?

   (1) $5\frac{1}{3}$ inches

   (2) $5\frac{2}{3}$ inches

   (3) $6\frac{1}{3}$ inches

   (4) $8\frac{2}{3}$ inches

   (5) Not enough information is given.

7. Jane ran $3\frac{3}{8}$ miles on Tuesday, $1\frac{1}{4}$ miles on Wednesday, and $2\frac{1}{2}$ miles on Thursday. How many total miles did Jane run?

   (1) $6\frac{7}{8}$

   (2) $7\frac{1}{8}$

   (3) $7\frac{7}{8}$

   (4) $8\frac{1}{8}$

   (5) $8\frac{7}{8}$

8. Tom wanted to combine boxes of pizza. He had $1\frac{1}{3}$ pizza, $1\frac{1}{4}$ pizza and $\frac{5}{8}$ pizza. How many boxes of pizza did Tom have?

   (1) $3\frac{11}{24}$

   (2) $3\frac{1}{4}$

   (3) $3\frac{5}{24}$

   (4) $2\frac{17}{24}$

   (5) $2\frac{5}{24}$

Ratios and Proportions

① Learn the Skill

A **ratio** is a comparison of two numbers. You can write a ratio as a fraction, using the word *to*, or with a colon (:). A **proportion** is an equation with a ratio on each side. The ratios are equal. You can use proportions to solve problems involving equal ratios.

② Practice the Skill

To succeed on the GED Mathematics Test, you must understand the concepts of rate and ratio and how to solve for each. Read the examples and strategies below. Then answer the question that follows.

A A ratio is different from a fraction. The bottom or second number of a ratio does not necessarily represent a whole. Therefore, you do not need to rename improper fractions as mixed numbers. However, ratios should still be simplified.

B A **unit rate** is a ratio with the denominator of 1. It can be expressed using the word *per*.

C In a proportion, the cross products are equal. Use cross products to solve proportions. If one of the four terms is missing, cross-multiply and divide the product by the third number to find the missing number.

Ratio

Jonathan earns $10 in 1 hour.

The ratio of dollars earned to hours is $\frac{10}{①}$, 10:① , or 10 to 1. **A**

This also can be written as $10 per hour.

Proportion

$\frac{3}{4} = \frac{6}{8}$ **Cross products:** $\frac{4 \times 6 = 24}{3 \times 8 = 24}$

$\frac{9}{12} = \frac{3}{x}$ ⟶ $12 \times 3 = 36$
$36 \div 9 = 4$
$x = 4$

☑ **TEST-TAKING TIPS**

When you write a proportion to solve a problem, the terms in both ratios need to be written in the same order. In problem 1, the top numbers can represent gallons and the bottom numbers can represent cost.

1. Carleen bought 3 gallons of milk for $12. How much would 5 gallons of milk cost?

(1) $9
(2) $12
(3) $16
(4) $20
(5) $24

Directions: Choose the <u>one best answer</u> to each question.

2. A store sold 92 pairs of pants and 64 shirts. What is the ratio of the number of pants sold to the number of shirts sold?

   (1) 92:64
   (2) 16:23
   (3) 64:92
   (4) 23:16
   (5) 16:92

3. Amanda traveled 558 miles in 9 hours. What is the unit rate that describes her travel?

   (1) 52 miles per hour
   (2) 61 miles per hour
   (3) 62 miles per hour
   (4) 71 miles per hour
   (5) 72 miles per hour

4. Jill mixed 2 cups of sugar with 10 cups of water to make lemonade. What ratio of sugar to water did she use?

   (1) $\frac{1}{5}$

   (2) $\frac{2}{10}$

   (3) $\frac{5}{1}$

   (4) $\frac{10}{2}$

   (5) Not enough information is given.

5. Sarah can ride 4 miles in 20 minutes on her bike. How many miles can she bike in 120 minutes?

   (1) 12
   (2) 24
   (3) 48
   (4) 120
   (5) 480

6. The ratio of adults to children on a field trip is 2:7. If there are 14 adults on the trip, how many children are there?

   (1) 4
   (2) 5
   (3) 7
   (4) 28
   (5) 49

7. Sam averages 65 miles per hour on a road trip. How many hours will it take him to drive 260 miles?

   (1) 3
   (2) 4
   (3) 5
   (4) 6
   (5) 7

8. The ratio of cars to trucks at an auto dealership is $\frac{3}{2}$. If there are 144 cars at the dealership, how many trucks are there?

   (1) 288
   (2) 240
   (3) 216
   (4) 144
   (5) 96

9. The Jammers basketball team won 25 games and lost 5 games during their season. What was their ratio of wins to losses?

   (1) 4:1
   (2) 5:1
   (3) 6:1
   (4) 8:1
   (5) 25:5

10. The GED preparation class has a teacher-to-student ratio of 1:12. If there are 36 students in the class, how many teachers are present?

   (1) 2
   (2) 3
   (3) 4
   (4) 6
   (5) 12

Decimals

① Learn the Skill

A **decimal** is another way to write a fraction. It uses the base-ten place value system. You can compare and order decimals using place value. You also can round decimals. Decimals include place values such as tenths, hundredths, and thousandths. Decimals can represent amounts much smaller than 1.

② Practice the Skill

One tenth is $\frac{1}{10}$ written as a fraction, or 0.1 written as a decimal. One hundredth is $\frac{1}{100}$ as a fraction, or 0.01 as a decimal. Read the examples and strategies below. Then answer the question that follows.

A Whole numbers are to the left of the decimal point and the decimals are to the right. As you move left on the place value chart, each column becomes 10 times greater. As you move to the right of the decimal point, each place value becomes $\frac{1}{10}$ the value of the column to its left. The number *20.3981* would become *203.981* if multiplied by 10. It would become *2.03981* if divided by 10.

B If two numbers have the same number of decimal places, compare and order decimals like whole numbers. If the number of decimal places is different, insert zeros to fill the empty decimal places. If the numbers have whole numbers and a decimal, compare the whole numbers first.

PLACE VALUE CHART									
Thousands			**Units**			**Decimals**			
hundreds	tens	ones	hundreds	tens	ones	tenths	hundredths	thousandths	ten thousandths
		,		2	0 .	3	9	8	1

Multiply by 10 for each cell you move left

Divide by 10 for each cell you move right

Compare Decimals

Compare the following decimals by using the > or < signs.

0.285 > 0.231   14.359 ___ 14.374

0.458 ___ 0.559   17.117 ___ 17.329

0.340 ___ 0.298   19.939 ___ 19.914

12.46 ___ 11.59   20.3981 ___ 20.3994

☑ **TEST-TAKING TIPS**

Circle the digit you want to round. If the digit to its right is greater than or equal to 5, add 1 to the circled digit. If it is less than 5, do not change it. Change the digits to the right of the rounded digit to zeros.

1. On her income tax form, Val rounds her income to the nearest dollar. If her income is $30,237.59, what will she write on her income tax form?

(1) $30,237.00
(2) $30,237.50
(3) $30,237.60
(4) $30,238.00
(5) $30,240.00

③ Apply the Skill

Directions: Choose the <u>one best answer</u> to each question.

2. For a science experiment, Ryan measures the distance a ball travels when rolled down a ramp. His measurement is 3.27 meters. He needs to record his measurement to the nearest tenth of a meter. What measurement will he record?

   (1) 3 meters
   (2) 3.2 meters
   (3) 3.27 meters
   (4) 3.3 meters
   (5) 4 meters

3. Dorie buys supplies for a camping trip. The cost of each item is listed below.

ITEM	COST
Camp suds	$2.95
Dried fruit	$5.65
Sleeping mat	$83.58
Water container	$12.28

   About how much money does Dorie spend in all? Round to the nearest dollar.

   (1) $104
   (2) $111
   (3) $115
   (4) $116
   (5) $120

4. The Warriors softball team had five players competing for the league's batting title. Which player had the highest batting average?

PLAYER	BATTING AVERAGE
Marie	.3278
Ellen	.3292
Krysten	.3304
Jennifer	.3279
Marti	.3289

   (1) Marie
   (2) Ellen
   (3) Krysten
   (4) Jennifer
   (5) Marti

<u>Questions 5 and 6</u> refer to the information below.

Sliced deli meat is sold by the pound. Shana bought five different meats at the deli.

DELI MEAT	WEIGHT
Chicken	1.59 pounds
Turkey	2.07 pounds
Ham	1.76 pounds
Salami	2.48 pounds
Roast beef	2.15 pounds

5. Which package of deli meat weighed the least?

   (1) chicken
   (2) turkey
   (3) ham
   (4) salami
   (5) roast beef

6. How many packages of deli meat weighed less than 2.25 pounds?

   (1) 1
   (2) 2
   (3) 3
   (4) 4
   (5) 5

7. Colin is packing the car with three pieces of luggage. He packs the heaviest piece into the car first, then the next heaviest, and finally the lightest piece on top. The weights of the three pieces of luggage are 25.57 pounds, 24.30 pounds, and 25.98 pounds. According to the weights, in which order will he pack the luggage into the car?

   (1) 24.30 pounds, 25.57 pounds, 25.98 pounds
   (2) 25.98 pounds, 24.30 pounds, 25.57 pounds
   (3) 25.98 pounds, 25.57 pounds, 24.30 pounds
   (4) 25.57 pounds, 24.30 pounds, 25.98 pounds
   (5) 24.30 pounds, 25.57 pounds, 24.30 pounds

Operations with Decimals

① Learn the Skill

You can add, subtract, multiply, and divide decimal numbers. When you perform operations with decimals, you must pay close attention to the placement of the decimal point. For example, when you add or subtract, write the numbers so that the place values and decimal points align.

② Practice the Skill

Problems on the GED Mathematics Test will require you to perform operations with decimals. Operations with decimals are slightly different than operations with whole numbers. You must understand these differences to correctly solve problems with decimals. Read the examples and strategies below. Then answer the question that follows.

A Align the decimal points. Then add or subtract like you do with whole numbers.

Addition

$$
\begin{array}{r}
{}^{1}\\
3.284\\
+\ 5.681\\
\hline
8.965
\end{array}
$$

Subtraction

$$
\begin{array}{r}
{}^{17}\\
{}^{5\ 7\ 11}\\
25.681\\
-\ 3.284\\
\hline
22.397
\end{array}
$$

Division

C

$$
\begin{array}{r}
12.283\\
8\overline{)98.264}\\
-8\\
\hline
18\\
-16\\
\hline
22\\
-16\\
\hline
66\\
-64\\
\hline
24
\end{array}
$$

C When dividing by a decimal, first make the divisor a whole number by moving the decimal place to the right. Then move the decimal point the same number of places to the right in the dividend. Add zeros to the dividend if necessary. For example,

0.08)12 ⟶ 8)1200

B Multiply as you do with whole numbers. Then count the decimal places in the factors. The number of decimal places in the product will be the sum of these two. Count from the right of the product to place the decimal point.

Multiplication

$$
\begin{array}{r}
5.61 \leftarrow \text{2 decimal places}\\
\times\ \ 3.8 \leftarrow \text{1 decimal place}\\
\hline
4488\\
+\ 16830\\
\hline
21.318 \leftarrow \text{3 decimal places}
\end{array}
$$

✓ TEST-TAKING TIPS

To multiply by 10, move the decimal one place to the right. To divide by 100, move the decimal two places to the left. The number of zeros shows how many spaces to move.

1. Molly bought coffee for $2.95 and a muffin for $1.29. She paid with a $5 bill. How much change did she receive?

   (1) $0.76
   (2) $0.86
   (3) $2.05
   (4) $4.14
   (5) $4.24

Directions: Choose the one best answer to each question.

Questions 2 through 4 refer to the following information:

Ben purchased groceries at Food U Eat. His receipt is shown below.

FOOD U EAT		
QUANTITY	ITEM	UNIT PRICE
5	Cereal	$3.85
6	Milk	$3.50
2	Butter	$1.29
4	Bread	$2.33

2. What amount did Ben pay for milk?

   (1) $14.00
   (2) $17.50
   (3) $18.00
   (4) $21.00
   (5) $23.10

3. How much more did Ben pay for the bread than for the butter?

   (1) $9.32
   (2) $6.84
   (3) $6.74
   (4) $3.62
   (5) $3.00

4. If Ben had purchased only milk and cereal, what would have been the amount of his bill?

   (1) $80.85
   (2) $40.25
   (3) $40.00
   (4) $19.25
   (5) $7.35

5. Anton bought four bowls for $33.40. If each bowl cost the same amount, what was the price of one bowl?

   (1) $5.56
   (2) $6.68
   (3) $8.35
   (4) $11.13
   (5) $16.70

6. Paper Plus sells reams of paper for $5.25 each. Discount Paper sells the same reams of paper for $3.99 each. How much would you save by purchasing 15 reams of paper at Discount Paper instead of at Paper Plus?

   (1) $1.26
   (2) $18.90
   (3) $59.85
   (4) $78.75
   (5) $138.60

7. David has a piece of rope that is 14.4 meters in length. If he divides the rope into 4 equal pieces, how many meters long will each piece be?

   (1) 57.6
   (2) 18.4
   (3) 10.4
   (4) 3.6
   (5) 2.2

Questions 8 and 9 refer to the following information.

Coach Steve needed to purchase new soccer equipment for the upcoming season.

EQUIPMENT	PRICE	QUANTITY
Soccer ball	$12.95	6
Shin guards	$10.95	12 sets
Knee pads	$8.95	12 sets
Uniforms	$17.00	12 sets

8. How much will Coach Steve spend on uniforms and soccer balls?

   (1) $29.95
   (2) $47.95
   (3) $97.80
   (4) $211.77
   (5) $281.70

9. How much more will Coach Steve spend on shin guards than knee pads?

   (1) $2.00
   (2) $16.00
   (3) $24.00
   (4) $36.00
   (5) $48.00

Fractions, Decimals, and Percent

① Learn the Skill

As with fractions and decimals, **percents** show part of a whole. Recall that with fractions, a whole can be divided into any number of equal parts. With a decimal, the number of equal parts must be a power of 10. Percent always compares amounts to 100. The percent sign, %, means "out of 100."

② Practice the Skill

Understanding how to convert between fractions, decimals, and percent will help you to efficiently solve problems on the GED Mathematics Test. Read the examples and strategies below. Then complete the table of conversions. When you are finished, answer the question that follows.

A To write a decimal as a percent, multiply by 100. Move the decimal point two places to the right and write the percent sign. To write a percent as a decimal, do the opposite.

B To write a fraction or mixed number as a decimal, divide the numerator by the denominator. To write a fraction or a mixed number as a percent, write it as a decimal and multiply by 100.

$$\frac{1}{5} = 0.2; 0.2 \times 100 = 20\%$$

To write a percent as a fraction or a mixed number, drop the percent sign and write the number as the numerator and 100 as the denominator. Then reduce the fraction.

$$20\% = \frac{20}{100} = \frac{1}{5}$$

C To rename a decimal as a fraction, write the number (without its decimal point) as the numerator. What is the place value of the last decimal digit? Write this as the denominator.

$$0.2 = \frac{2}{10} = \frac{1}{5}$$

FRACTION	DECIMAL	PERCENT
$\frac{1}{5}$	0.2	20%
$\frac{1}{4}$	0.25	25%
$\frac{1}{2}$	0.5	50%
$\frac{3}{4}$	0.75	
	1.8	

TEST-TAKING TIPS

If a number is shown without a decimal point, such as 80%, assume that it lies directly to the right of the ones digit.

1. In a neighborhood, 27 of the 45 children are in elementary school. What percent of the children in the neighborhood are in elementary school?

   (1) 20%
   (2) 25%
   (3) 40%
   (4) 60%
   (5) 166%

Directions: Choose the <u>one best answer</u> to each question.

2. Shelly's Boutique is advertising 25% off all merchandise. What fraction of the original price will customers save during the sale?

(1) $\frac{2}{3}$

(2) $\frac{1}{2}$

(3) $\frac{1}{3}$

(4) $\frac{1}{4}$

(5) $\frac{1}{5}$

3. The unit price of a can of peas is $7\frac{1}{2}$ cents per ounce. To find the price of an 8-ounce can, Jeff first writes the mixed number as a decimal. What decimal does he write?

(1) 7.2
(2) 7.5
(3) 7.55
(4) 7.6
(5) 7.75

4. City Electric provides electricity for $\frac{3}{8}$ of the homes in Center City. For what percentage of the homes does City Electric provide electricity?

(1) 37.5%
(2) 37%
(3) 36.5%
(4) 36%
(5) 35.5%

5. In the spring, $\frac{1}{8}$ of the students at a community college participate in a work-study program. What decimal describes $\frac{1}{8}$?

(1) 0.08
(2) 0.105
(3) 0.125
(4) 0.8
(5) 8.0

6. In a survey, 0.22 of the respondents answered "Yes" to the question "Would you consider voting for a candidate from a third party?" What fraction of respondents answered "No"?

(1) $\frac{11}{50}$

(2) $\frac{22}{100}$

(3) $\frac{22}{50}$

(4) $\frac{39}{50}$

(5) $\frac{78}{10}$

7. The Strikers girls soccer team won 9 of its 13 games. What percentage of games did the Strikers win?

(1) 61.5%
(2) 66.7%
(3) 69.2%
(4) 76.9%
(5) 84.6%

8. On the GED Mathematics pretest, Jarrod answered 88% of the questions correctly. If there were 25 questions on the test, how many did Jarrod correctly answer?

(1) 18
(2) 19
(3) 20
(4) 21
(5) 22

9. At Bright Minds Learning, 75% of employees work as instructors. If there are 300 employees at Bright Minds Learning, how many of them work as instructors?

(1) 150
(2) 175
(3) 200
(4) 225
(5) 250

Percent Problems

① Learn the Skill

There are three main parts of a percent problem—the base, the rate, and the part. The **base** is the whole amount. The **rate** tells how the base and whole are related. The **part** is a piece of the whole or base. The rate is always followed by a percent sign. You can use proportions to solve percent problems. You can also use the percent formula: **base × rate = part**.

② Practice the Skill

You must understand how the base, part, and rate are related to successfully solve percent problems on the GED Mathematics Test. As you practice percent problems, many common percents will become familiar to you, and eventually you will be able to use mental math to solve the problems. Read the examples and strategies below. Then answer the question that follows.

A Set the rate over 100 to equal the part over the base.

B You may be asked to find the percent of change. Subtract to find the difference between the original and new amounts. Then divide the difference by the original amount. Convert the decimal to a percent.

C P = the amount of money borrowed, rate (r) is the percent charged, and time (t) is the time in years that you are borrowing money. Before you calculate, change the rate to a decimal and make sure the time is in years. Write months as a fraction or a decimal.

Use a Proportion
Zach answered 86% of the questions on a math exam correctly. If there were 50 questions, how many questions did Zach answer correctly?

$$\frac{\text{Part}}{\text{Base}} = \frac{\text{Rate}}{100} \qquad \frac{?}{50} = \frac{86}{100}$$

$$50 \times 86 = 4300 \longrightarrow 4300 \div 100 = \textbf{43 questions}$$

Find Percent Increase or Decrease
Last year, Kareem paid $750 per month for rent. This year he pays $820 a month. What is the percent of increase?

$$\begin{cases} \$820 - \$750 = \$70.00 \\ \$70.00 \div \$750 = 0.09 \\ 0.09 \times 100 = \textbf{9\%} \end{cases}$$

Interest Problems **C**
Kelly took out a $20,000 loan for four years at 3% interest. How much interest (I) will she pay on the loan?

$$I = prt$$
$$I = \$20,000 \times 0.03 \times 4$$
$$I = \textbf{\$2,400}$$

☑ TEST-TAKING TIPS

When using the formula **base × rate = part**, write the rate as a decimal.

1. Kirsten borrowed $1,000 from her sister for six months. If she pays 5% interest, what is the total that she will owe her sister in six months?

   (1) $1,000
   (2) $1,025
   (3) $1,050
   (4) $1,075
   (5) $1,300

③ Apply the Skill 🖩

Directions: Choose the <u>one best answer</u> to each question.

2. Tia earns $552 per week. Of this amount, 12% is deducted for taxes. What amount is deducted each week?

   (1) $6.62
   (2) $55.20
   (3) $66.00
   (4) $66.24
   (5) $485.76

3. At a food packaging factory, 309 of the 824 employees work the third shift. What percentage of employees work the third shift?

   (1) 25%
   (2) 32.5%
   (3) 35%
   (4) 37.5%
   (5) 40%

4. Andrew received a raise from $24,580.00 per year to $25,317.40 per year. What percent raise did he receive?

   (1) 1%
   (2) 2%
   (3) 3%
   (4) 4%
   (5) 5%

5. Isabella paid $425 for a new bicycle, plus 6% sales tax. What is the total amount she paid?

   (1) $25.50
   (2) $27.50
   (3) $427.50
   (4) $450.50
   (5) $457.50

6. A computer company received 420 customer service calls in one day. Forty-five percent of the calls were about software issues. How many of the calls were about software?

   (1) 19
   (2) 189
   (3) 210
   (4) 229
   (5) 231

7. A furniture store is having a sale. A sofa is regularly priced at $659 but is on sale for 20% off. What is the sale price of the sofa?

   (1) $649.00
   (2) $527.20
   (3) $450.80
   (4) $427.50
   (5) $131.80

8. Daria invested $5,000 in an account that earns 5% interest over nine months. How much interest will she earn over that time?

   (1) $5,187.50
   (2) $3,750.00
   (3) $2,250.00
   (4) $250.00
   (5) $187.50

9. Fredrica's take-home pay is $2,250 per month. She spends 20% of this on her rent. How much does Fredrica spend each month on rent?

   (1) $400
   (2) $450
   (3) $500
   (4) $550
   (5) $600

10. At a publishing house, Herb supervises 25 employees. His staff will grow by 40% next month. How many employees will Herb have on staff next month?

    (1) 35
    (2) 45
    (3) 55
    (4) 65
    (5) 75

Unit 1 Review

On the GED Mathematics Test you will be asked to write your answers in different ways. Below are two ways to write your answers for this Unit Review.

Horizontal-response format

①②●④⑤

To record your answers, fill in the numbered circle that corresponds to the answer you select for each question in the Unit Review. Do not rest your pencil on the answer area while considering your answer. Make no stray or unnecessary marks. If you change an answer, erase your first mark completely. Mark only one answer space for each question; multiple answers will be scored as incorrect.

Alternate-response format

To record your answers for an alternate format question
- Begin in any column that will allow your answer to be entered;
- Write your answer in the boxes in the top row;
- In the column beneath a fraction bar or decimal point (if any) and each number in your answer, fill in the bubble representing that character;
- Leave blank any unused column.

Directions: Choose the one best answer to each question.

1. Thirty-five percent of residents surveyed were in favor of creating a new road. The remaining residents objected. If 1,200 people were surveyed, how many objected to the new road?

   (1) 35
   (2) 360
   (3) 420
   (4) 600
   (5) 780

   ①②③④⑤

2. Dina purchased a new dining room table for $764.50 and four new chairs for $65.30 each. What was the cost of the whole set?

   (1) $699.20
   (2) $829.80
   (3) $926.30
   (4) $1,025.70
   (5) $1,091.00

   ①②③④⑤

3. The Martins drove 210.5 miles on the first day of their trip and 135.8 miles the second day. How many more miles did they drive the first day than the second day?

   (1) 60.0
   (2) 74.7
   (3) 149.4
   (4) 271.6
   (5) 346.3

   ①②③④⑤

4. Erin must add $4\frac{1}{2}$ cups of flour to her cookie batter using a $1\frac{1}{2}$-cup measuring cup. How many times will she need to fill the measuring cup with flour?

   (1) one
   (2) two
   (3) three
   (4) four
   (5) five

   ①②③④⑤

Questions 5 and 6 refer to the following information and table.

The table shows the breakdown of after-school options for students at Oak Ridge Elementary School.

WHAT STUDENTS DO AFTER SCHOOL	
OPTION	NUMBER OF STUDENTS
Parent pickup	118
Walk	54
Bus	468
After-school programs	224

5. What fraction of the students walk home?

(1) $\frac{1}{216}$

(2) $\frac{9}{216}$

(3) $\frac{1}{24}$

(4) $\frac{1}{16}$

(5) $\frac{1}{3}$

①②③④⑤

6. What fraction of the students take the bus or stay after school?

(1) $\frac{197}{432}$

(2) $\frac{468}{864}$

(3) $\frac{117}{216}$

(4) $\frac{173}{216}$

(5) $\frac{13}{24}$

①②③④⑤

7. Kara invested $1,250 in the production of a friend's music CD. Her friend paid her back at 6% annual interest after 36 months. How much money did Kara get back?

(1) $225
(2) $500
(3) $1,250
(4) $1,025
(5) $1,475

①②③④⑤

8. Ken needs a cable that is $4\frac{3}{4}$ meters long. He has a cable that is $5\frac{1}{3}$ meters long. What fraction of a meter will Ken need to cut off?

```
     [ ] [/] [/] [/] [ ]
     [.] [.] [.] [.] [.]
     [0] [0] [0] [0] [0]
     [1] [1] [1] [1] [1]
     [2] [2] [2] [2] [2]
     [3] [3] [3] [3] [3]
     [4] [4] [4] [4] [4]
     [5] [5] [5] [5] [5]
     [6] [6] [6] [6] [6]
     [7] [7] [7] [7] [7]
     [8] [8] [8] [8] [8]
     [9] [9] [9] [9] [9]
```

9. Tracy bought two pretzels for $1.95 each and two soft drinks for $0.99 each. If she paid with a $10 bill, how much change did she receive?

(1) $4.12
(2) $5.12
(3) $6.10
(4) $7.06
(5) $8.02

①②③④⑤

Questions 10 and 11 refer to the following information and table.

A number of women participate in five different intramural college sports. The fraction of women who participate in each sport is shown in the table.

WOMEN'S INTRAMURAL SPORTS	
SPORT	FRACTION OF WOMEN
Basketball	$\frac{1}{6}$
Volleyball	$\frac{1}{20}$
Soccer	$\frac{1}{3}$
Ultimate frisbee	$\frac{1}{5}$
Lacrosse	$\frac{1}{4}$

10. In which sport do the greatest number of women participate?

(1) basketball
(2) volleyball
(3) soccer
(4) ultimate frisbee
(5) lacrosse

①②③④⑤

11. What fraction of women participates in lacrosse and basketball?

(1) $\frac{2}{12}$

(2) $\frac{2}{10}$

(3) $\frac{5}{12}$

(4) $\frac{5}{8}$

(5) $\frac{3}{4}$

①②③④⑤

12. In a class of 408 students, 28 students have last names that begin with the letter *M*. What fraction of students is this?

13. Benjamin drove a distance of 301.5 miles in 4.5 hours. If Benjamin drove at a constant rate, how many miles per hour did he drive?

(1) 63
(2) 64
(3) 65
(4) 66
(5) 67

①②③④⑤

14. Scarlett purchased 20 shares of AD stock at $43 per share. She sold the 20 shares at $52 per share. How much money did Scarlett make on her investment?

(1) $80
(2) $180
(3) $200
(4) $280
(5) $860

①②③④⑤

15. A group of 426 people is going to a rally. Each bus can take 65 people. What is the minimum number of buses needed?

(1) 4
(2) 5
(3) 6
(4) 7
(5) 8

①②③④⑤

16. Alice typed her income into tax-preparation software. If her income was fifty-six thousand, two hundred, twenty-eight dollars, what digits did she type?

(1) 5, 6, 2, 2, 0, 8
(2) 5, 0, 6, 2, 2, 8
(3) 5, 6, 2, 2, 8
(4) 5, 6, 2, 0, 8
(5) 5, 6, 2, 8

①②③④⑤

17. Delaney has $198 in her checking account. She deposits $246 and writes checks for $54 and $92. How much is left in her account?

Questions 18 through 20 refer to the following information and table.

Kurt and his family went to the state fair. They ate lunch at a wild game restaurant. The menu is shown below.

ITEM	PRICE
Walleye fillet	$5.89
Elk sandwich	$9.65
Wild boar barbecue	$9.19
Salmon on a stick	$5.45
Kid's buffalo platter	$3.50

18. What is the most expensive item on the menu?

(1) Walleye fillet
(2) Elk sandwich
(3) Wild boar barbecue
(4) Salmon on a stick
(5) Kid's buffalo platter

①②③④⑤

19. Kurt ordered 1 wild boar barbecue, 1 walleye fillet, and 3 kid's buffalo platters. If he brought $50 with him to the fair, how much does he have left?

(1) $18.58
(2) $24.42
(3) $25.58
(4) $26.42
(5) $31.42

①②③④⑤

20. How much more do 2 elk sandwiches cost than 3 kid's platters?

(1) $2.39
(2) $6.15
(3) $7.88
(4) $8.80
(5) $15.80

①②③④⑤

21. The ratio of men to women in a chorus is 2:3. If there are 180 women in the chorus, how many men are in the chorus?

(1) 80
(2) 100
(3) 120
(4) 160
(5) 180

①②③④⑤

22. Anna can knit a scarf in $1\frac{2}{3}$ hours. How many scarves can she knit in 4 hours?

(1) $2\frac{2}{5}$

(2) $2\frac{2}{3}$

(3) $3\frac{1}{5}$

(4) $3\frac{2}{3}$

(5) $3\frac{3}{5}$

①②③④⑤

23. Eighty-four percent of student athletes attended a preseason meeting. If there are 175 student athletes, how many attended the meeting?

(1) 84
(2) 128
(3) 137
(4) 147
(5) 149

①②③④⑤

24. Two-thirds of Mrs. Jensen's class passed the science exam. If there are 24 students in her class, how many passed the exam?

(1) 12
(2) 13
(3) 14
(4) 15
(5) 16

①②③④⑤

Questions 25 and 26 refer to the following information and table.

During an election year, 200 people were surveyed about their political affiliation. The results are shown in the table.

VOTERS' POLL	
PARTY AFFILIATION	NUMBER OF PEOPLE
Democratic	78
Republican	64
Independent	46
Green	10
Libertarian	2

25. What is the ratio of Green party supporters to Libertarian party supporters?

(1) 5 to 1
(2) 10 to 1
(3) 1 to 10
(4) 1 to 5
(5) 2 to 10

①②③④⑤

26. If 400 people were surveyed, how many would you expect to affiliate themselves with the Democratic Party?

(1) 278
(2) 156
(3) 78
(4) 39
(5) Not enough information is given.

①②③④⑤

27. The population of a city grew from 43,209 to 45,687 in just five years. What was the percent increase in the population to the nearest whole percent?

28. Fifty-four percent of customers at a grocery store bought milk on Friday. What fraction of the customers is this?

(1) $\frac{27}{50}$

(2) $\frac{14}{25}$

(3) $\frac{8}{17}$

(4) $\frac{3}{5}$

(5) Not enough information is given.

①②③④⑤

29. Rodrigo pays $165.40 per month on his car loan. How much does he pay on his loan in 1 year?

(1) $661.60
(2) $719.13
(3) $992.40
(4) $1,984.80
(5) $3,969.60

①②③④⑤

30. A muffin recipe calls for $1\frac{3}{8}$ cups of oil. If Sean triples the recipe, how many cups of oil does he need?

(1) $3\frac{1}{8}$

(2) $3\frac{3}{8}$

(3) $4\frac{1}{8}$

(4) $4\frac{1}{4}$

(5) $4\frac{3}{8}$

①②③④⑤

31. A certain type of cheese sells for $8.99 per pound. What is the cost of a 1.76-pound block of cheese?

(1) $5.10
(2) $12.82
(3) $14.38
(4) $15.80
(5) $15.82

①②③④⑤

Question 32 refers to the following table.

MILES BIKED	
Jackson	26.375
Ben	$25\frac{4}{5}$
Stefan	32.95

32. How many more miles did Stefan ride than Ben?

(1) 7.0

(2) 7.15

(3) $7\frac{3}{5}$

(4) 7.25

(5) $7\frac{4}{5}$

①②③④⑤

Unit 2

CHRISTOPHER BLIZZARD

Christopher Blizzard develops open-source software that enables people to explore the Internet.

Christopher Blizzard likes to cast a wide—and free—'Net. As a leading expert in the math-heavy world of Web-based software development, Blizzard believes that the Internet should remain a public resource accessible by all. To that end, Blizzard, who earned his GED certificate in 1994, helps develop non-proprietary software that allows people to explore the Internet.

Blizzard works for one of the leading providers of open-source software. Open-source software (OSS) products, such as the Internet browser Mozilla Firefox, are built and maintained by a network of volunteer programmers. Blizzard has been a longtime contributor to open-source projects, most notably for Mozilla and the One Laptop per Child (OLPC) project.

The purpose of One Laptop per Child is to demonstrate the effectiveness of laptops as learning tools for children in developing countries. While serving as team leader and designing elements of the OLPC system, Blizzard focused on this important question:

> **"How do we create an interesting social environment for kids to share and learn together?"**

As part of his job, Blizzard learned a variety of computer programming languages. He and others have been involved in the ongoing development of the Firefox browser, which launched in 2005 and has been downloaded more than 60 million times worldwide. He also sat on the Board of Directors of the Mozilla Foundation.

BIO BLAST: Christopher Blizzard

- Works for one of the leading providers of open-source software
- Contributes to open-source projects, such as One Laptop per Child
- Worked previously as systems engineer and software developer
- Served on Mozilla Foundation's Board of Directors

Measurement/Data Analysis

Unit 2: Measurement/Data Analysis

Each time you step on a scale, plan a trip, or cook a meal, you are using skills related to measurement and data analysis. The growing use of computers and the Internet has assisted in the collection, storage, and interpretation of large sets of data. Such information often is presented in graphs.

Skills used to measure and analyze data are important both to your everyday life as well as to your success on the GED Mathematics Test. As with other subject areas, measurement and data analysis comprise between 20 and 30 percent of mathematics questions on the GED Mathematics Test. In Unit 2, you will study different measurement systems and forms of measurement, along with probability, time, and visual ways of displaying data. Such skills will help you prepare for the GED Mathematics Test.

Table of Contents

Measurement Systems and Units of Measure

1 Learn the Skill

When solving measurement problems, you will use either the **U. S. customary system** or the **metric system**. **Units of measure** in the U. S. customary system include inch and foot (length), ounce and pound (weight), and pint and quart (capacity). Units of measure in the metric system include centimeter and meter (length), gram and kilogram (mass), and milliliter and liter (capacity).

2 Practice the Skill

By mastering the skill of using measurement systems and converting among units of measure within each system, you will improve your study and test-taking skills, especially as they relate to the GED Mathematics Test. Read the tables and strategies below. Then answer the question that follows.

U.S. CUSTOMARY UNITS OF MEASURE

Length	**Liquid Capacity**	**Weight**
1 foot (ft) = 12 inches (in.)	1 cup (c) = 8 fluid ounces (fl oz)	1 pound (lb) = 16 ounces (oz)
1 yard (yd) = 3 feet	1 pint (pt) = 2 cups	1 ton (tn) = 2,000 pounds
1 mile (mi) = 5,280 feet	1 quart (qt) = 2 pints	
1 mile = 1,760 yards	1 gallon (gal) = 4 quarts	

METRIC UNITS OF MEASURE

Length	**Capacity**
1 kilometer (km) = 1,000 meters (m)	1 kiloliter (kL) = 1,000 liters (L)
1 meter (m) = 100 centimeters (cm)	1 liter (L) = 100 centiliters (cL)
1 centimeter (cm) = 10 millimeters (mm)	1 centiliter (cL) = 10 milliliters (mL)

Mass
1 kilogram (kg) = 1,000 grams (g)
1 gram (g) = 100 centigrams (cg)
1 centigram (cg) = 10 milligrams (mg)

A When you convert and rename a unit in the metric system, multiply or divide by 10, 100, or 1,000. The following prefixes can help in making metric conversions.

milli- means 1/1000	*deca-* means 10
centi- means 1/100	*hecto-* means 100
deci- means 1/10	*kilo-* means 1,000

☑ TEST-TAKING TIPS

First identify the measurement system and the units of measure that are being converted. If you are converting from a lesser unit to a greater unit, divide. If you are converting from a greater unit to a lesser unit, multiply.

1. Dante mixes 30 milliliters of one liquid with 2 centiliters of a second liquid. How many centiliters of liquid does he have altogether?

(1) 5 cL
(2) 32 cL
(3) 50 cL
(4) 302 cL
(5) 500 cL

<u>Directions</u>: Choose the <u>one best answer</u> to each question.

Use your own knowledge and the tables on p. 30 to convert and rename the units of measure.

2. Samantha is building a miniature maze for a science experiment. She determines that she needs 6 yards of wood for the exterior walls and 12 feet of the same wood for the interior walls. How many feet of wood must she buy in order to build her maze?

    (1) 10 ft
    (2) 18 ft
    (3) 24 ft
    (4) 30 ft
    (5) 54 ft

3. Over a two-day track meet, Jason ran in one 2-kilometer race, two 1,500-meter races, and five 100-meter races. How many meters did Jason run over the two days?

    (1) 3,500 m
    (2) 4,000 m
    (3) 4,500 m
    (4) 5,000 m
    (5) 5,500 m

4. Mr. Trask wants to fill his four hummingbird feeders with liquid food. Two feeders hold 6 fluid ounces each. One larger feeder holds 1 cup of liquid. The largest feeder holds 1 pint. How many fluid ounces of liquid food does Mr. Trask need to fill the four bird feeders?

    (1) 8 fl oz
    (2) 14 fl oz
    (3) 28 fl oz
    (4) 30 fl oz
    (5) 36 fl oz

Questions 5 through 7 refer to the table below.

Five students in Ms. Craig's chemistry class conducted an experiment using different amounts of the same powdered chemicals.

CHEMICAL AMOUNTS FOR EXPERIMENTS			
STUDENT	CHEMICAL A	CHEMICAL B	CHEMICAL C
Tory	2 cg	5 cg	2 g
Shantell	5 mg	8 cg	60 cg
Diego	15 mg	50 cg	1 g
Janice	15 mg	2 cg	50 cg
Dana	5 mg	3 cg	0.5 g

5. How many centigrams of Chemical A were used by the five students?

    (1) 6 cg
    (2) 15 cg
    (3) 42 cg
    (4) 60 cg
    (5) 402 cg

6. How much greater was the mass of Chemical C than the mass of Chemical A in Shantell's experiment?

    (1) 10 mg
    (2) 55 mg
    (3) 100 mg
    (4) 550 mg
    (5) 595 mg

7. Dana used a total of 535 mg of chemicals for her experiment. How much greater was the mass of chemicals used by Diego?

    (1) 10 mg
    (2) 170 mg
    (3) 980 mg
    (4) 1,765 mg
    (5) 8,965 mg

UNIT 2

LESSON 2
Length, Perimeter, and Circumference

① Learn the Skill

The distance around a polygon, such as a triangle or rectangle, is called the **perimeter**. Determine the perimeter of a polygon by measuring and adding the **lengths** of all its sides. The perimeter of a circle is called the **circumference**. To use the formula for circumference, you must know the diameter of a circle. A diameter runs across the center of a circle from one point on the circle to another. If you know the radius of a circle, double it to find the diameter.

② Practice the Skill

By mastering the skills of measuring lengths and finding both perimeter and circumference, you will improve your study and test-taking skills, especially as they relate to the GED Mathematics Test. Read the example and strategies below. Then answer the question that follows.

A Identify the important information given in the paragraph and in the figure. The paragraph tells you that the hotel workers decide to make the diameter of the fence *twice* that of the pool. The figure gives you the diameter of the pool. You need to know both pieces of information to answer the question.

B Using the formula for circumference requires that you know either the length of the **diameter** (a chord that passes through the center of a circle) or the **radius** (any line segment from the center of the circle to a point on the circle). The radius is always half the length of the diameter. This figure shows the diameter of the pool.

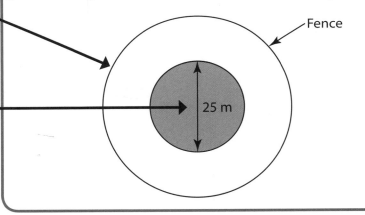

Workers at the Vista Hotel want to erect a circular fence around a circular swimming pool. They decide to make the diameter of the fence twice that of the pool. The workers need to know the circumference of the fence in order to buy the correct amount of metal fencing.

Fence

25 m

☑ TEST-TAKING TIPS

When multiplying by *pi*, use estimation to help you choose a reasonable answer. To multiply 50 by 3.14, first round 3.14 to 3. Since 50 × 3 = 150, answer choice 3 seems to be the most reasonable.

1. What is the circumference of the fence that will surround the Vista Hotel swimming pool?

(1) 50 m
(2) 78.5 m
(3) 157 m
(4) 785 m
(5) Not enough information is given.

32   Lesson 2 | Length, Perimeter, and Circumference

Directions: Choose the <u>one best answer</u> to each question.

<u>Questions 2 through 5</u> refer to the following figures.

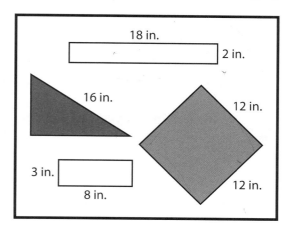

2. What is the perimeter of the red square?

   (1) 12 in.
   (2) 24 in.
   (3) 48 in.
   (4) 144 in.
   (5) Not enough information is given.

3. What is the perimeter of the long white rectangle?

   (1) 18 in.
   (2) 20 in.
   (3) 36 in.
   (4) 38 in.
   (5) 40 in.

4. What is the perimeter of the red triangle?

   (1) 16 in.
   (2) 34 in.
   (3) 40 in.
   (4) 48 in.
   (5) Not enough information given.

5. What is the perimeter of the short white rectangle?

   (1) 11 in.
   (2) 20 in.
   (3) 21 in.
   (4) 22 in.
   (5) 24 in.

<u>Questions 6 through 8</u> refer to the following figure.

A home-building company wants to divide a large parcel of land into three separate parcels. Parcel B and Parcel C will have the same perimeter. The parcel of land is shown below.

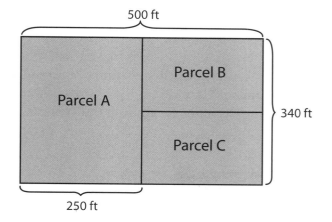

6. What is the perimeter of Parcel A?

   (1) 500 ft
   (2) 840 ft
   (3) 1,000 ft
   (4) 1,090 ft
   (5) 1,180 ft

7. What is the perimeter of Parcel C?

   (1) 420 ft
   (2) 590 ft
   (3) 840 ft
   (4) 1,340 ft
   (5) 1,680 ft

8. What effect would the addition of Parcel D have on the overall perimeter of the land?

   (1) It would increase.
   (2) It would decrease.
   (3) It would remain the same.
   (4) It first would increase and then decrease.
   (5) It first would decrease and then increase.

9. Mary used about 56.5 cm of yarn to form a circle. Which could be the diameter of the circle?

   (1) 9.0 cm
   (2) 18.0 cm
   (3) 88.7 cm
   (4) 177.4 cm
   (5) Not enough information is given.

UNIT 2

Area

① Learn the Skill

Area is the amount of space that covers a two-dimensional figure. Square units are used to measure area. In this lesson, you will find the area of rectangles, triangles, parallelograms, and trapezoids. A **parallelogram** is a four-sided figure with opposite pairs of congruent and parallel sides. A **trapezoid** is a four-sided figure with only one pair of parallel sides.

The formulas for finding the area may require you to identify the **base** and **height** of a figure. The base and height form a right angle. A trapezoid has two bases. To find the bases of a trapezoid, look for the two parallel sides.

② Practice the Skill

By mastering the skill of finding area, you will practice skills that involve using formulas and simplifying expressions. Read the example and strategies below. Then answer the question that follows.

A Since the door is square, you know that each side is 12 ft. The length and width of a square are equal. The formula for the area of a square is $A = s \times s$, where s represents the length of one side.

B The units used to express the area of a figure are based on the unit of length. In this problem, the length is measured in feet, so the area is measured in square feet (sq ft or ft^2).

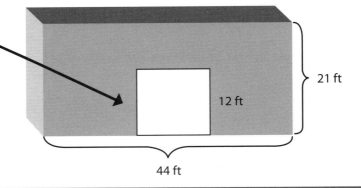

Jorge is in charge of painting the front wall of a storage facility. The facility has a large square metal door that does not need to be painted.

21 ft

12 ft

44 ft

☑ TEST-TAKING TIPS

Some word problems require you to complete several steps to find the answer. Finding the area of the square door in this problem is not sufficient, nor is finding the area of the wall including the door. To solve the problem, you must find the area of the wall and the area of the door. Then you must perform a third step.

1. What is the area of the front wall of the storage facility that Jorge must paint?

(1) 144 sq ft
(2) 288 sq ft
(3) 672 sq ft **B**
(4) 780 sq ft
(5) 924 sq ft

Directions: Choose the <u>one best answer</u> to each question.

<u>Questions 2 through 4</u> refer to the following information and figure.

Melanie is having new hardwood floors installed in her bedroom and closet. Below is a diagram of her bedroom and closet, which are both rectangular.

MELANIE'S BEDROOM

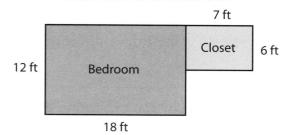

2. How much flooring will Melanie need to purchase to cover her bedroom floor?

    (1) 26 ft²
    (2) 42 ft²
    (3) 60 ft²
    (4) 216 ft²
    (5) 324 ft²

3. Each square foot of flooring costs $6. How much will it cost to cover the bedroom and closet?

    (1) $516
    (2) $648
    (3) $1,152
    (4) $1,374
    (5) $1,548

4. How much would it cost if Melanie chose only to install hardwood flooring in her bedroom?

    (1) $252
    (2) $504
    (3) $756
    (4) $1,296
    (5) $1,548

<u>Question 5</u> refers to the following information and figure.

Vanessa designed the logo for a neighborhood walkathon. She is coloring every other strip in her drawing with red ink. Each strip has the same area.

WALK FOR OUR PARK

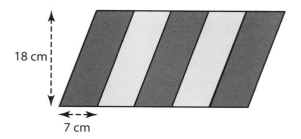

5. What is the area of the logo that Vanessa wants to cover with red ink?

    (1) 25 sq cm
    (2) 75 sq cm
    (3) 126 sq cm
    (4) 378 sq cm
    (5) 630 sq cm

<u>Question 6</u> refers to the following figure.

Quentin built a triangular pen for his dog Toro. The pen is shown below.

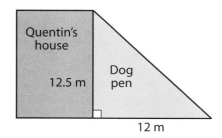

6. What is the area in which Toro can run?

    (1) 75 sq m
    (2) 150 sq m
    (3) 162.5 sq m
    (4) 600.25 sq m
    (5) Not enough information is given.

UNIT 2

Volume

① Learn the Skill

Three-dimensional figures have **volume**, which is the amount of space that exists inside a figure. A **rectangular prism** is a very common three-dimensional figure. It is shaped like a box. A **cube** is a special kind of rectangular prism because it has 6 congruent square faces.

Volume is measured in cubic units. The units used to measure the volume are based on the units used to measure the dimensions of the figure. For example, if the length, width, and height of a rectangular prism are measured in centimeters, the volume is expressed in cubic centimeters, or cm^3. The GED Mathematics Test provides formulas for finding the volume of these three-dimensional figures.

② Practice the Skill

By mastering the skill of finding volume, you will improve your study and test-taking skills, especially as they relate to the GED Mathematics Test. Read the example and strategies below. Then answer the question that follows.

A The paragraph says that the clerk wants to know about the amount of water inside a three-dimensional figure. This means that the clerk wants to know the *volume*. Knowing precisely what a problem is asking you to do is the first step toward correctly solving it.

B In problems that require you to find volume, the next step is to identify the kind of three-dimensional figure being used in the problem. The tank has the shape of a rectangular prism, so the volume formula for rectangular prisms is needed. You should become familiar with the volume formula for cubes and rectangular prisms.

Julia went to a pet store to buy a new filter for her aquarium. The clerk wanted to know the amount of water in the fish tank.

JULIA'S FISH TANK

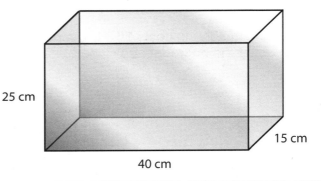

25 cm

15 cm

40 cm

☑ TEST-TAKING TIPS

Since multiplication is commutative, you can multiply numbers in any order. It may be simpler to multiply the product of 25 and 40 by 15. To multiply 25 by 40, think of $25 \times 4 \times 10$.

1. What is the volume of Julia's fish tank?

(1) 80 cm^3
(2) 600 cm^3
(3) 1,000 cm^3
(4) 3,125 cm^3
(5) 15,000 cm^3

Directions: Choose the <u>one best answer</u> to each question.

<u>Questions 2 through 4</u> refer to the following three-dimensional figures.

The concession stand at a baseball stadium sells popcorn at the same price but in two different containers. Manny wants to determine which is the better buy.

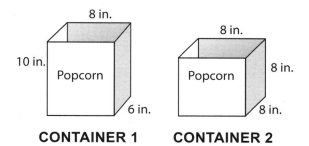

CONTAINER 1   CONTAINER 2

2. If Manny wants to know about the amount in each container, which measurement should he calculate?

(1) perimeter
(2) volume
(3) circumference
(4) area
(5) Not enough information is given.

3. What is the volume of Container 1?

(1) 48 in.³
(2) 80 in.³
(3) 60 in.³
(4) 480 in.³
(5) 576 in.³

4. What is the volume of Container 2?

(1) 24 in.³
(2) 32 in.³
(3) 64 in.³
(4) 128 in.³
(5) 512 in.³

<u>Question 5</u> refers to the figure below.

Contractors at a state fair need to know the volume of a certain building so that they can install adequate air conditioner units.

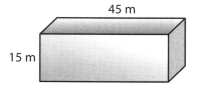

5. What is the volume of the building that will need to be air-conditioned?

(1) 60 m³
(2) 675 m³
(3) 10,125 m³
(4) 21,600 m³
(5) Not enough information is given.

<u>Question 6</u> refers to the following information and figure.

A carpenter plans to build a floor-to-ceiling closet in a large loft room. The height of the room is 12 ft. The closet is the section shown on the right.

CARPENTER'S FLOORPLAN

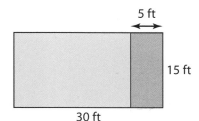

6. By how much will the volume of the loft decrease after the closet is built?

(1) 375 ft³
(2) 450 ft³
(3) 900 ft³
(4) 4500 ft³
(5) 5400 ft³

Mean, Median, and Mode

① Learn the Skill

The mean, median, mode, and range are values used to describe a set of data. The **mean** is the average value of a data set. The **median** is the middle number in a set of data when the values are ordered from least to greatest. In the number set 23, 24, 27, 29, 75, the median is 27. Notice that the median was not affected by the value of 75, which is much larger than the other values. As a result, the median more accurately describes the set than the mean (35.6) does. The **mode** is the value that occurs most frequently in a set of data. To find the **range**, subtract the least value from the greatest value.

② Practice the Skill

By mastering the skills of finding mean, median, mode, and range, you will study data in a meaningful way. Read the example and strategies below. Then answer the question that follows.

A To find the median of a data set, list the values in order from least to greatest. The number 65 is listed three times in the table. When ordering numbers, be sure to list 65 three times.

B When a number set consists of an odd number of values, the middle number is the median. When the set consists of an even number of data points, find the mean of the two middle numbers. Note that the median may not be a number in the set of data.

Felipe measured and recorded the heights of the runners participating in a neighborhood relay race.

HEIGHTS OF RELAY RACE RUNNERS	
RUNNER	**HEIGHT (INCHES)**
Carol	63
Steven	68
Pedro	65 **A**
Julia	65 **A**
Chantell	67
Camille	64
Frank	72
William	71
Jane	65 **A**
Jake	72

☑ TEST-TAKING TIPS

There are ten values in the table. When listing values from least to greatest, check that you listed a total of ten values.

1. What is the median height of the runners?

(1) 65 in.
(2) 65.2 in.
(3) 66 in.
(4) 67 in.
(5) 67.2 in.

UNIT 2

Directions: Choose the <u>one best answer</u> to each question.

<u>Questions 2 through 4</u> refer to the following information and table.

The running times of a YMCA-sponsored 100-meter race are shown below.

TIMES FOR THE 100-METER RACE	
RUNNER	**TIME (SECONDS)**
David	13.5
Sanya	16.0
Jeremy	12.6
Erica	15.2
Chen	12.8
Yusuf	11.8
Matt	17.2
Sarah	12.1

2. What is the range of the runners' times in the 100-meter race?

    (1) 4.2 s
    (2) 5.4 s
    (3) 6.4 s
    (4) 11.8 s
    (5) 13.9 s

3. What is the median time in the race?

    (1) 5.4 s
    (2) 12 s
    (3) 13.15 s
    (4) 13.9 s
    (5) 16 s

4. What is the difference between Sarah's time and the mean time of the runners?

    (1) 1.35 s
    (2) 1.8 s
    (3) 13.9 s
    (4) 13.15 s
    (5) Not enough information is given.

<u>Question 5</u> refers to the following table.

The owner of Ice Cream Palace listed the number of milk shakes sold each day for one week.

DAILY MILK SHAKE SALES						
DAY	**MON.**	**TUES.**	**WED.**	**THURS.**	**FRI.**	**SAT.**
Milk Shakes Sold	22	16	20	26	24	85

5. Which value best describes the number of milk shakes sold at Ice Cream Palace on a typical summer day?

    (1) 20
    (2) 21.6
    (3) 23
    (4) 32.16
    (5) 69

<u>Question 6</u> refers to the following table.

The sneaker sales at Sneaker World were recorded each day for one week.

SNEAKER WORLD SALES	
DAY	**TOTAL SALES**
Monday	$5,229
Tuesday	$3,598
Wednesday	$6,055
Thursday	$3,110
Friday	$3,765
Saturday	?

6. The mean sales for this one week were $3,743.14. The manager misplaced her records for Saturday. What were the sales on Saturday?

    (1) $634.99
    (2) $701.84
    (3) $852.19
    (4) $975.92
    (5) $1,681.15

Probability

① Learn the Skill

When you flip a quarter, you have an equal chance of flipping heads or tails. The chances of heads can be expressed as *1:2*, where *1* represents the number of favored outcomes (flipping heads) and *2* represents the number of possible outcomes. This ratio expresses the **theoretical probability** of the event. In theory, each time you flip a coin, you have a 50% chance of flipping heads.

Probability based on the results of an experiment is called **experimental probability**. As with theoretical probability, you can express experimental probability as a ratio, fraction, or percent. If you toss a quarter ten times and get heads six times, the experimental probability is $\frac{6}{10}$, which simplifies to $\frac{3}{5}$.

② Practice the Skill

By mastering the skill of probability, you will improve your study and test-taking skills, especially as they relate to the GED Mathematics Test. Read the example and strategies below. Then answer the question that follows.

A By choosing a striped marble from the bag during the first event and not replacing it, Marc affected the outcome of the second event. The two events are said to be **dependent**. When events are dependent, the number of outcomes changes.

If Marc had replaced the marble after the first event, the first event would not have affected the outcome of the second event. In this case, the first event and the second event would have been **independent**.

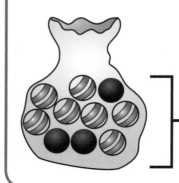

A bag of 10 marbles contains 7 striped marbles and 3 black marbles.

B Probability can be expressed as a ratio. If the bag contained two black marbles and three striped marbles, the probability of drawing a black marble during the first event would be 2:5. There are two black marbles and five possible outcomes. The same probability can be expressed as a fraction ($\frac{2}{5}$), a decimal (0.4), and a percentage (40%).

✓ TEST-TAKING TIPS

When answering a probability problem, always check whether the events are independent or dependent. Then determine the probability in the form that is easiest for you.

1. In the first event, Marc draws a striped marble. <u>He does not replace it</u>. In the next three events, Marc draws 2 striped marbles and 1 black marble. He does not replace the marbles. What is the probability that he will select a black marble on the fifth event?

   (1) 1:10
   (2) 1:3
   (3) 2:7
   (4) 2:3
   (5) 2:2

Directions: Choose the <u>one best answer</u> to each question.

Questions 2 through 4 refer to the following spinner.

Maude uses this spinner to conduct a probability experiment.

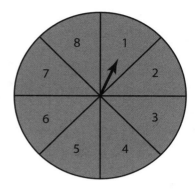

2. On the first spin, what is the probability that the spinner will land on 6?

   (1) 1:8
   (2) 1:7
   (3) 1:6
   (4) 6:8
   (5) Not enough information is given.

3. On the second spin, what is the probability that the spinner will land on 4 or 8?

   (1) 0.48
   (2) 0.28
   (3) 0.25
   (4) 0.16
   (5) 0.13

4. Maude spins the spinner twice. She lands on 4 and 6. So far, what is her experimental probability of spinning an odd number?

   (1) $\frac{1}{6}$

   (2) $\frac{1}{8}$

   (3) $\frac{1}{2}$

   (4) $\frac{1}{1}$

   (5) $\frac{0}{2}$

Questions 5 and 6 refer to the table below.

A large chain store keeps track of its daily customer complaint calls.

COMPLAINT CALLS	
DEPARTMENT	**NUMBER OF COMPLAINTS**
Electronics	6
Housewares	4
Automotive	2
Clothing	3

5. What is the probability that the next complaint call to the store will concern the clothing department?

   (1) 20%
   (2) 25%
   (3) 30%
   (4) 50%
   (5) Not enough information is given.

6. What is the probability that the next complaint call will concern the electronics department or the housewares department?

   (1) $\frac{4}{15}$

   (2) $\frac{1}{2}$

   (3) $\frac{3}{5}$

   (4) $\frac{2}{3}$

   (5) $\frac{4}{5}$

7. Ian read in the newspaper that there is a 40% chance of rain tomorrow. What is the probability that it will not rain tomorrow?

   (1) $\frac{1}{25}$

   (2) $\frac{3}{50}$

   (3) $\frac{3}{5}$

   (4) $\frac{2}{3}$

   (5) $\frac{3}{2}$

UNIT 2

Tables

① Learn the Skill

As you learned on p. vii, and practiced in parts of Units 1 and 2, **tables** organize data so that information is easier to identify and compare. On the GED Mathematics Test, you will be asked to interpret and compare data from different kinds of tables. These might include frequency tables, which are explained below.

Tables can contain more information than you want or need in solving a problem. When taking the GED Mathematics Test, carefully read questions involving tables to identify precisely the information that you need in solving a problem.

② Practice the Skill

By mastering the skill of using tables, you will improve your study and test-taking skills, especially as they relate to the GED Mathematics Test. Read the example and strategies below. Then answer the question that follows.

A The number of data values for given intervals is known as the **frequency**. A frequency table presents the exact number of data values for a given interval. This frequency table displays five intervals.

B The table enables you to quickly eliminate certain answer choices. You can see at a glance that the data for the first and last frequencies show the least amount in the table. The sharp decline in frequency for the 46–55 age group also helps you eliminate that interval as a possible answer choice.

Advertisers are interested in knowing when people of various ages watch television. Such information helps advertisers know which products to advertise at certain times of the day.

AGE GROUP	NIGHTTIME TELEVISION VIEWERS BY AGE GROUP	
	FREQUENCY	
	7 P.M.–9 P.M.	9 P.M.–11 P.M.
16–25	355	790
26–35	1,047	1,532
36–45	1,212	1,519
46–55	1,357	399
56–65	887	352

☑ TEST-TAKING TIPS

Before answering questions about tables, be sure to read the title and column headings. They contain important information that will help you understand what the data represent.

1. Which age group would advertisers most likely want to target between the hours of 7 P.M. and 11 P.M.?

    (1) 16–25
    (2) 26–35
    (3) 36–45
    (4) 46–55
    (5) 56–65

 **Apply the Skill**

Directions: Choose the <u>one best answer</u> to each question.

<u>Questions 2 and 3</u> refer to the information and table below.

In one week, Ms. Cappelli drives from her hometown of Bakerton to several towns in her region.

MILES BETWEEN CITIES					
MILEAGE CHART	**ALBAN**	**ASHLAND**	**BAKERTON**	**BENTON**	**CLARK**
Alban	—	139	302	79	221
Ashland	139	—	83	115	169
Bakerton	302	83	—	274	111
Benton	79	115	274	—	203
Clark	221	169	111	203	—

2. Ms. Cappelli drives from Bakerton to Benton on Monday. On Tuesday she drives from Benton to Clark. How many miles has she driven by the time she arrives in Clark?

    (1) 71
    (2) 163
    (3) 314
    (4) 385
    (5) 477

3. On Wednesday, Ms. Cappelli drives from Clark to Ashland for an appointment. She then drives home to spend the night. How many miles does she drive on Wednesday?

    (1) 194
    (2) 252
    (3) 280
    (4) 304
    (5) 338

<u>Questions 4 and 5</u> refer to the following table.

A newspaper reporter keeps track of the average price of gasoline in his county for eight weeks.

AVERAGE COST OF GASOLINE IN MAY AND JUNE	
WEEK	**AVERAGE COST PER GALLON**
1	$2.98
2	$2.89
3	$3.24
4	$3.09
5	$3.45
6	$3.62
7	$3.04
8	$2.99

4. Between which two weeks did the average price of a gallon of gasoline increase by the greatest amount?

    (1) Week 1–Week 2
    (2) Week 2–Week 3
    (3) Week 3–Week 4
    (4) Week 4–Week 5
    (5) Week 5–Week 6

5. Which statement is true about the average price of gasoline per gallon during Week 2?

    (1) It represents the greatest decrease in price from one week to the next.
    (2) It is the last time the average price fell from one week to another.
    (3) It is the lowest average price per gallon during the eight-week period.
    (4) The price increase from Week 2 to Week 3 is the sharpest increase shown.
    (5) It is the last time in the eight-week period that the average price was under three dollars.

UNIT 2

Bar and Line Graphs

① Learn the Skill

Graphs are used to visually display data. **Bar graphs** use vertical or horizontal bars to show data. These graphs are typically used to compare data. **Line graphs** are best suited for showing how a data set changes over time. Graphs may include scales that provide detail about the data.

Scatter plots are a type of line graph that show how one set of data affects another. The relationship between data sets is known as its **correlation.** A correlation may be positive (extending upward from the origin to x- and y- points) or negative (extending downward from the y-axis to the x-axis), or it may not exist at all.

② Practice the Skill

By mastering the skill of interpreting bar and line graphs, you will improve your study and test-taking skills, especially as they relate to the GED Mathematics Test. Read the example and strategies below. Then answer the question that follows.

Ⓐ Multiple sets of data can appear on a bar graph or a line graph. When it occurs in a line graph, such as this one, you will see two or more line patterns. The lines for each park are a different color.

Ⓑ When using a graph, first examine its different parts. The title describes the topic of the graph. Labels along the vertical and horizontal axes describe the data. The scale of the vertical axis shows the interval being used. You will find categories along the horizontal axis. This line graph also has a key that shows the color code used for the two different parks.

This line graph shows the monthly rainfall through the spring and summer at two state parks.

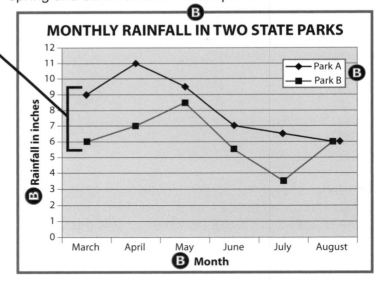

TEST-TAKING TIPS

Be sure you know precisely what the question is asking. For example, you can tell at a glance that the months May and August are not realistic answer choices.

1. During which month was the difference in rainfall between the two parks the greatest?

    (1) March
    (2) April
    (3) May
    (4) June
    (5) July

Directions: Choose the <u>one best answer</u> to each question.

<u>Questions 2 through 4</u> refer to the bar graph below.

Fred records the long jump results in a track meet. He creates the following bar graph to show the results online.

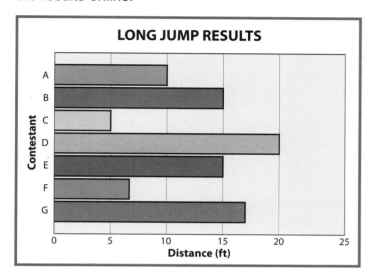

2. Which contestant jumped exactly half as far as the contest winner?

   (1) A
   (2) B
   (3) C
   (4) D
   (5) E

3. Katie and Alana jumped the same distance. How far did they each jump?

   (1) 5 ft
   (2) 10 ft
   (3) 15 ft
   (4) 17 ft
   (5) 20 ft

4. What is the range of scores from the event?

   (1) 10
   (2) 12.5
   (3) 12.7
   (4) 13
   (5) 15

<u>Questions 5 and 6</u> refer to the scatter plot.

An educational services company compared student scores on the GED Mathematics Test with the amount of hours they prepared for it. Their findings are shown in the scatter plot below.

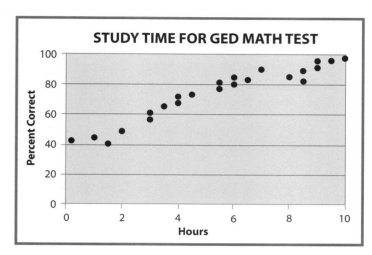

5. Anton hopes to earn an 80% or higher on the GED Mathematics Test. How many hours should he study?

   (1) 0
   (2) 2
   (3) 4
   (4) 6
   (5) 8

6. Which statement does the data imply?

   (1) There is no correlation between hours spent studying and scores on the GED Mathematics Test.
   (2) A student is more likely to score higher on the GED Mathematics Test if he/she spends more hours studying.
   (3) A student cannot score well on the GED Mathematics Test if he/she only studies for 4 hours.
   (4) A student is less likely to score higher on the GED Mathematics Test if he/she spends more hours studying.
   (5) Not enough information is given.

Circle Graphs

① Learn the Skill

Like bar and line graphs, circle graphs show data visually. Whereas a line graph shows how data changes over time, a **circle graph** shows how parts compare to a whole. A circle graph of sales from each department in a store can show at a glance the most productive department, as well as how each department's sales compares to that of the whole store.

Values of the circle graph sections may be expressed as percents, decimals, fractions, or whole numbers. You may need to convert from one form to another.

② Practice the Skill

By mastering the skill of interpreting circle graphs, you will improve your study and test-taking skills, as well as your understanding of percents, decimals, and fractions. Read the example and strategies below. Then answer the question that follows.

A Some circle graphs, such as this one, are labeled with only categories, rather than categories and percent. In whatever manner a circle graph is labeled, the whole circle represents 1, or 100%.

B Use a category's size to estimate its value. Notice that car maintenance and gasoline each represent about a quarter of the whole, or 25%. This helps you estimate the percentages of other categories.

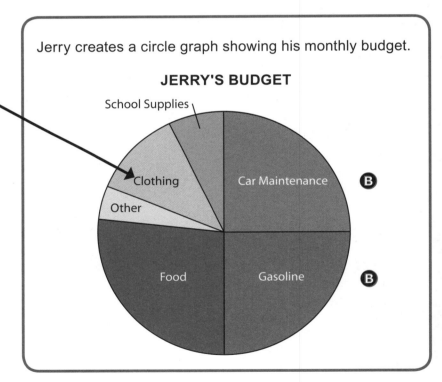

Jerry creates a circle graph showing his monthly budget.

JERRY'S BUDGET

✓ TEST-TAKING TIPS

Since the section for food is larger than the section for car maintenance or gasoline, you can estimate that Jerry budgets more than 25% for food.

1. Approximately what percentage per month does Jerry budget for food?

   (1) 10%
   (2) 20%
   (3) 25%
   (4) 30%
   (5) 45%

UNIT 2

Directions: Choose the <u>one best answer</u> to each question.

Questions 2 and 3 refer to the following circle graph.

The circle graph below shows the methods of transportation that employees use to get to work.

HOW EMPLOYEES GET TO WORK

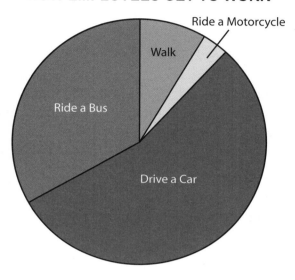

2. About what part of the employee population drives a car to work?

   (1) 20%
   (2) 25%
   (3) 30%
   (4) 50%
   (5) 60%

3. Which of the following events would likely result in an increase in the amount of people who walk or ride a bus to work each day?

   (1) a decrease in road construction
   (2) a sharp rise in gasoline prices
   (3) greater affordability of hybrid vehicles
   (4) lower prices from motorcycle manufacturers
   (5) wearing down of infrastructure, such as sidewalks

Questions 4 through 6 refer to the following circle graph.

A library creates a circle graph of the types of books checked out by readers in September.

WHAT PEOPLE READ IN SEPTEMBER

4. Which category of books has fewer readers than crime books?

   (1) mystery
   (2) classic fiction
   (3) romance
   (4) nonfiction
   (5) poetry

5. Which categories of books could a librarian make the best argument to order in August?

   (1) nonfiction and mystery
   (2) mystery and romance
   (3) nonfiction and crime
   (4) romance and crime
   (5) classic fiction and poetry

6. If 30,000 books were checked out in September, approximately how many were romance?

   (1) 18,000
   (2) 15,000
   (3) 7,500
   (4) 4,500
   (5) 2,000

Time

1 Learn the Skill

On the GED Mathematics Test, you will be asked to answer questions involving **time**. Some questions will require that you determine **elapsed time**, or the amount of time that has passed from one time to another. For example, if a baseball game starts at 1:15 P.M. and ends at 4:35 P.M., you can calculate the elapsed time of the baseball game as follows:

From 1:15 to 2:00 is 45 minutes. From 2:00 to 4:35 is 2 hours 35 minutes.

The length of the game is 45 minutes + 2 hours + 35 minutes, or 2 hours 80 minutes.

Rename 80 minutes as 1 hour 20 minutes.

2 hours + 1 hour 20 minutes = 3 hours 20 minutes, so the game lasted 3 hours and 20 minutes.

Other questions may involve the formula *distance = rate × time*. If you know any two of the values in the formula, you can solve for the third value.

2 Practice the Skill

By mastering the skill of calculating time, you will improve your study and test-taking skills, especially as they relate to the GED Mathematics Test. Read the example and strategies below. Then answer the question that follows.

A In this problem, the text gives you important data that you need to solve the problem. By substituting 155 for distance and 65 for rate in the formula, you can see that you need to divide 155 by 65 to find the time, 2.38 hours. You can round this value to 2.4 hours.

> Cheryl is driving to her mother's house. She is traveling at an average speed of <u>65 mph</u>. She has <u>155 miles</u> left before she reaches her destination. Glancing at her watch, she sees that it is 1:00 P.M. Cheryl uses the following formula to help her figure out the approximate time of arrival at her mother's house:
>
> *distance = rate × time*

B To answer the question, convert 0.4 h to minutes by multiplying by 60 (the number of minutes in an hour). When calculating the approximate time of Cheryl's arrival, remember that the time is added from 1:00 P.M.

☑ TEST-TAKING TIPS

When using the formula *distance = rate × time*, pay close attention to the units. For example, if you are calculating the rate in miles per hour, the units you use for time must be in hours.

1. About what time can Cheryl expect to reach her mother's house, providing that she maintains her average speed and does not stop?

   (1) 2:38 P.M.
   (2) 3:00 P.M.
   (3) 3:25 P.M.
   (4) 3:45 P.M.
   (5) 5:00 P.M.

Directions: Choose the one best answer to each question.

Questions 2 and 3 refer to the following schedule.

TRAINS FROM CENTRAL STATION		
TIME	DESTINATION	GATE
11:52 A.M.	Ridley	31
1:05 P.M.	Harrison	28
2:35 P.M.	Ridley	32
3:20 P.M.	Harrison	29

2. If a passenger misses the 11:52 train to Ridley, how long must he or she wait for the next train to Ridley?

   (1) 1 hr 5 min
   (2) 1 hr 13 min
   (3) 2 hr 35 min
   (4) 2 hr 43 min
   (5) 3 hr 15 min

3. The journey to Harrison takes 1 hour 45 minutes. If Lance takes the 3:20 train, what time can he expect to arrive at Harrison?

   (1) 1:45 P.M.
   (2) 4:45 P.M.
   (3) 5:05 P.M.
   (4) 5:45 P.M.
   (5) Not enough information is given.

4. Mindy finished a race in 34.4 minutes. Sarah finished the same race in 34.9 minutes. How much faster was Mindy's time?

   (1) 0.5 s
   (2) 5.0 s
   (3) 12.0 s
   (4) 30.0 s
   (5) 50.0 s

Questions 5 and 6 refer to the following paragraph.

An express bus leaves Cleveland, Ohio, at 11:35 P.M. and travels 462 miles to New York City. It arrives in New York City at 7:05 A.M.

5. How long did the bus ride take?

   (1) 7 hr 0 min
   (2) 7 hr 30 min
   (3) 7 hr 40 min
   (4) 8 hr 0 min
   (5) 8 hr 15 min

6. At what average speed, in miles per hour, did the bus travel?

   (1) 64.3
   (2) 63.3
   (3) 62.2
   (4) 61.6
   (5) 60.8

Question 7 refers to the following table.

The table shows the top three runners' times in a recent marathon.

MARATHON TIMES	
RUNNER	TIME
Sasha Singleton	3 hr 16 min 6 s
Niabi Greene	3 hr 20 min 10 s
Carol Carlisle	3 hr 22 min 2 s

7. How much faster was the winning time than the second-fastest time?

   (1) 0 min 4 s
   (2) 3 min 4 s
   (3) 3 min 56 s
   (4) 4 min 0 s
   (5) 4 min 4 s

UNIT 2

Unit 2 Review

On the GED Mathematics Test you will be asked to write your answers in different ways. Below are two ways to write your answers for this Unit Review.

Horizontal-response format

① ② ● ④ ⑤

To record your answers, fill in the numbered circle that corresponds to the answer you select for each question in the Unit Review. Do not rest your pencil on the answer area while considering your answer. Make no stray or unnecessary marks. If you change an answer, erase your first mark completely. Mark only one answer space for each question; multiple answers will be scored as incorrect.

Alternate-response format

To record your answers for an alternate format question
- Begin in any column that will allow your answer to be entered;
- Write your answer in the boxes in the top row;
- In the column beneath a fraction bar or decimal point (if any) and each number in your answer, fill in the bubble representing that character;
- Leave blank any unused column.

Directions: Choose the <u>one best answer</u> to each question.

Question 1 refers to the information below.

> **Length**
> 1 foot (ft) = 12 inches (in.)
> 1 yard (yd) = 3 feet
> 1 mile (mi) = 5,280 feet
> 1 mile = 1,760 yards

1. An artist needs 840 feet of ribbon for an outdoor work of art. The company that manufactures the ribbon only sells it in large rolls of 100 yards. How many yards of ribbon does the artist want to use?

   (1) 70 yd
   (2) 280 yd
   (3) 300 yd
   (4) 350 yd
   (5) 2,520 yd

Question 2 refers to the information below.

> **Mass**
> 1 kilogram (kg) = 1,000 grams (g)
> 1 gram (g) = 100 centigrams (cg)
> 1 centigram = 10 milligrams (mg)

2. A shoelace has a mass of 1 gram. A textbook has a mass of about 1 kilogram. How many shoelaces would you need to gather to equal the mass of two textbooks?

   (1) 100
   (2) 200
   (3) 1,000
   (4) 2,000
   (5) Not enough information is given.

① ② ③ ④ ⑤

① ② ③ ④ ⑤

Question 3 refers to the figure below.

3. A gardener is planting flowers along the edge of two triangular sections of a large garden. Both triangular plots are the same size.

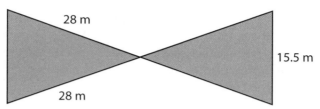

What is the perimeter of the two triangular garden plots?

(1) 56 m
(2) 71.5 m
(3) 127.5 m
(4) 143 m
(5) Not enough information is given.

 ①②③④⑤

4. Bill knows that the diameter of the circular sign is 30 inches. What is the circumference in inches?

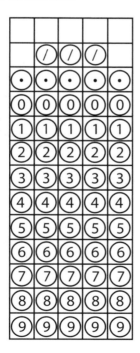

Questions 5 through 7 refer to the following text and figure.

An architect designs a dividing wall with three moveable sections for a conference room at a large hotel. The two triangular shapes are the same size.

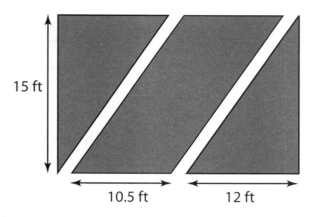

5. What is the area of the middle parallelogram?

(1) 37.5 ft²
(2) 110.25 ft²
(3) 150.0 ft²
(4) 157.5 ft²
(5) Not enough information is given.

①②③④⑤

6. What is the area of one of the triangles?

(1) 37.5 ft²
(2) 72.0 ft²
(3) 90.0 ft²
(4) 112.5 ft²
(5) 180.0 ft²

①②③④⑤

7. When the three pieces fit together, what is the area of the dividing wall?

(1) 75.0 ft²
(2) 247.5 ft²
(3) 337.5 ft²
(4) 517.5 ft²
(5) Not enough information is given.

 ①②③④⑤

Questions 8 and 9 refer to the following information and table.

A cable television company asks a family to keep track of the number of hours they spend watching television. They record weekly data for two months.

WEEKLY TELEVISION VIEWING	
WEEK	HOURS WATCHED
1	21.5
2	28.0
3	15.5
4	23.0
5	29.0
6	34.0
7	27.0
8	35.0

8. To the nearest hundredth, what is the mean number of hours watched each week by the family?

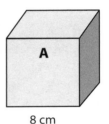

9. Which statement best describes the mean and median?

(1) The median is slightly greater than the mean.
(2) The median and mean are equal.
(3) The median is significantly greater than the mean.
(4) The mean is slightly greater than the median.
(5) The mean is significantly greater than the median.

①②③④⑤

Questions 10 and 11 refer to the following figures.

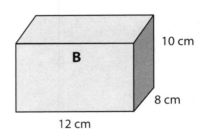

10. Container A is a cube. How many cubic centimeters could fill Container A?

(1) 8 cm³
(2) 16 cm³
(3) 64 cm³
(4) 128 cm³
(5) 512 cm³

①②③④⑤

11. Henry fills the two containers with water. How much water will he use?

(1) 1,472 cm³
(2) 1,044 cm³
(3) 1,024 cm³
(4) 976 cm³
(5) 968 cm³

①②③④⑤

Questions 12 and 13 refer to the following text and figure.

The die has one of the digits 1 through 6 on each side.

12. What is the probability, expressed as a fraction, of rolling a 2 or a 4?

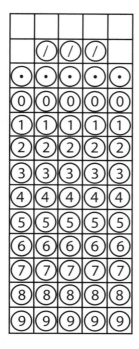

13. What is the probability of rolling an even number?

(1) 25%
(2) 33%
(3) 50%
(4) 66%
(5) 100%

①②③④⑤

14. Jane is making an ornament that consists of powders of different colors. She wants 250 grams of blue powder, 250 grams of silver powder, 300 grams of red powder, and 375 grams of green powder. How many kilograms of powder will Jane need altogether?

(1) 1.175 kg
(2) 11.75 kg
(3) 117.5 kg
(4) 1,175 kg
(5) 11,750 kg

①②③④⑤

Question 15 refers to the following text and figure.

Contractors installing a wire fence around two tennis courts need to find the perimeter so that they know how much fence to order. Each tennis court is 60 feet wide and 120 feet long.

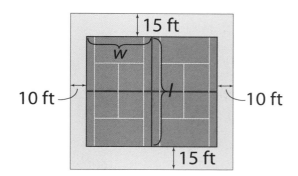

15. How much fencing do the contractors need to order to enclose the tennis courts?

(1) 290 ft
(2) 360 ft
(3) 520 ft
(4) 530 ft
(5) 580 ft

①②③④⑤

Questions 16 and 17 refer to the following text and line graph.

A company keeps track of the bonuses its employees receive each year.

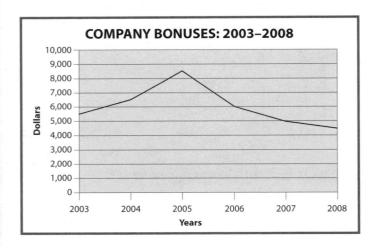

COMPANY BONUSES: 2003–2008

16. Between which two years did the amount of bonuses awarded show the greatest increase?

(1) 2003–2004
(2) 2004–2005
(3) 2005–2006
(4) 2006–2007
(5) 2007–2008

①②③④⑤

17. During which year was the amount of bonuses awarded less than $5,000?

(1) 2003
(2) 2005
(3) 2006
(4) 2007
(5) 2008

①②③④⑤

18. Devaughn averages only 45 mph driving along a mountain road. How many miles, expressed as a decimal, can Devaughn travel in 45 minutes?

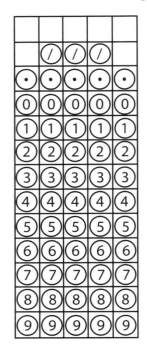

19. The flight distance between Boston and Chicago is approximately 850 miles. A commercial airliner leaves Boston at 11:30 A.M. The average speed of the plane is 500 mph. The time in Boston is one hour ahead of the time in Chicago. What time will it be in Chicago when the plane lands?

(1) 11:12 A.M.
(2) 12:12 P.M.
(3) 1:07 P.M.
(4) 1:42 P.M.
(5) Not enough information is given.

①②③④⑤

20. Every evening, Mrs. Jackson walks around her neighborhood. The walk usually takes her about 25 minutes. What is the distance of her walk?

(1) 0.25 mi
(2) 0.5 mi
(3) 1.0 mi
(4) 1.4 mi
(5) Not enough information is given.

①②③④⑤

Question 21 refers to the following text and diagram.

Kim wants to paint a circular dance floor.

21. Which measurement should Kim calculate to determine the amount of paint he should purchase?

(1) perimeter
(2) area
(3) circumference
(4) volume
(5) length

①②③④⑤

22. A cardboard box has the dimensions shown below.

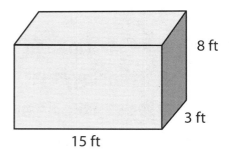

8 ft

3 ft

15 ft

If the length, width, and height were each doubled, what would be the new volume?

(1) 2,880 ft³
(2) 960 ft³
(3) 480 ft³
(4) 104 ft³
(5) 52 ft³

①②③④⑤

23. The most popular game at a carnival is *Spin for Fortune*. The "Sorry" outcome means that the player does not win a prize.

What are the chances, expressed as a decimal, that a player spinning the wheel will not win a prize?

	/	/	/	
⊙	⊙	⊙	⊙	⊙
⓪	⓪	⓪	⓪	⓪
①	①	①	①	①
②	②	②	②	②
③	③	③	③	③
④	④	④	④	④
⑤	⑤	⑤	⑤	⑤
⑥	⑥	⑥	⑥	⑥
⑦	⑦	⑦	⑦	⑦
⑧	⑧	⑧	⑧	⑧
⑨	⑨	⑨	⑨	⑨

24. A hockey team traveling in a bus left one school at 11:50 A.M. They arrived at another school at 2:10 P.M. How long was the trip?

(1) 1 hr 50 min
(2) 2 hr 10 min
(3) 2 hr 20 min
(4) 2 hr 30 min
(5) 3 hr 10 min

①②③④⑤

Unit 3

PHILIP EMEAGWALI

Philip Emeagwali is known as the "Bill Gates of Africa." It's easy to see why. As with Gates, the famed founder of Microsoft, Emeagwali left school before receiving his high school diploma. Like Gates, Emeagwali, a native of Nigeria, has enjoyed tremendous success in the computer industry.

Emeagwali left school when his family no longer could afford to pay for his education. He instead taught himself subjects such as mathematics, physics, chemistry, and English. Such efforts enabled Emeagwali to pass the General Certificate of Education exam (the British version of the GED Tests) and earn a scholarship to Oregon State University. It was only after Emeagwali arrived in the United States that he first used a telephone, visited a library, or saw a computer. Emeagwali graduated from Oregon State and went on to earn master's degrees in civil engineering, marine engineering, and mathematics.

Emeagwali's determination has served as a springboard to his success in the field of supercomputing. In 1989, a computer system that he built became the first to perform 3.1 billion calculations per second. Emeagwali used this computer to help scientists understand how oil flowed underground. For his efforts, he was awarded the prestigious Gordon Bell Prize, considered the Nobel Prize of computing. As he notes,

Philip Emeagwali used his success on the British version of the GED Tests as a springboard to multiple college degrees and a career in the field of supercomputing.

> **" I find supercomputing to be a fascinating, challenging, and critical technology that can be used to solve many societal problems, such as predicting the spread of AIDS ... "**

BIO BLAST: Philip Emeagwali

- Invented a program for the Connection Machine, the fastest computer on Earth

- Designed a system of parallel computers used by search engines such as Yahoo

- Developed the Hyperball computer, which can forecast global-warming patterns

- Conducted research to help solve problems in the areas of meteorology, energy, health, and the environment

Algebra, Functions, and Patterns

Unit 3: Algebra, Functions, and Patterns

Algebra builds upon the core areas of mathematics, such as number sense and data measurement and analysis, by translating everyday situations into mathematical language. We use algebra to solve complex problems and to explore more sophisticated areas of mathematics. Certain jobs, such as those in high-tech fields, require strong backgrounds in algebra and other forms of higher mathematics.

As with other subject areas, algebra, functions, and patterns make up between 20 and 30 percent of questions on the GED Mathematics Test. In Unit 3, you will study integers, equations, patterns, factoring, and other skills that will help you prepare for the GED Mathematics Test.

Table of Contents

Integers

① Learn the Skill

Integers include positive whole numbers (1, 2, 3, . . .), their opposites or negative numbers (–1, –2, –3, . . .), and zero. Positive numbers show an increase and may be written with or without a plus sign. Negative numbers show a decrease and are written with a negative sign. Integers can be added, subtracted, multiplied, and divided. There are specific rules for adding, subtracting, multiplying, and dividing integers.

② Practice the Skill

Many mathematics problems relating to real-world situations use integers. You must understand and follow the rules for adding, subtracting, multiplying, and dividing integers to solve these problems on the GED Mathematics Test. Read the examples and strategies below. Then answer the question that follows.

A If integers have like signs, add and keep the common sign. If integers have different signs, find the difference. The sign is negative if you started with more negative, or positive if you started with more positive.

B To subtract an integer, add its opposite. For example, the opposite of +5 is –5.

C For multiplying or dividing integers: If the signs are the same, the answer will be positive. If the signs are different, the answer will be negative.

Add Integers

$(+4) + (+7) = +11$ $(-5) + (-9) = -14$
$(-8) + (+4) = -4$ $(-5) + (+12) = +7$

Subtract Integers

$(+8) - (-5) = (+8) + (+5) = 13$
$8 - 5 = 8 + (-5) = 3$

Multiply and Divide Integers

$(4)(5) = +20$ $(-4)(5) = -20$
$(-4)(-5) = 20$ $(4)(-5) = -20$

$18 \div 9 = 2$ $(-18) \div (9) = -2$
$(-18) \div (-9) = 2$ $18 \div (-9) = -2$

☑ TEST-TAKING TIPS

It may be helpful to use a number line when solving problems that involve integers. To solve 12 – (–3), begin at –3 and count spaces to 12. You will see that the distance is +15.

1. In the morning, the temperature was –3°F. By mid-afternoon, the temperature was 12°F. What was the change in temperature between the morning and afternoon?

    (1) –15°F
    (2) –9°F
    (3) 9°F
    (4) 12°F
    (5) 15°F

Directions: Choose the <u>one best answer</u> to each question.

2. In a board game, Dora moves forward 3 spaces, back 4 spaces, and forward again 8 spaces in one turn. What is her net gain or loss of spaces?

   (1) 1 space forward
   (2) 7 spaces forward
   (3) 1 space backward
   (4) 7 spaces backward
   (5) 15 spaces forward

3. Uyen has a balance of $154 in her savings account. She withdraws $40 from a cash machine. What is her new balance?

   (1) $94
   (2) $100
   (3) $104
   (4) $114
   (5) $194

4. Sasha's home is 212 feet above sea level. She participated in a scuba dive in which she descended to 80 feet below sea level. Which integer describes Sasha's change in position from her house to the lowest point of her dive?

   (1) −292
   (2) −132
   (3) 132
   (4) 292
   (5) 302

5. Tyler enters his office building on the ground floor. He walks 6 floors up to his office. He heads down 2 floors to the cafeteria, and after lunch he walks up 4 floors for a meeting. Where is Tyler's meeting?

   (1) the 4th floor
   (2) the 5th floor
   (3) the 6th floor
   (4) the 7th floor
   (5) the 8th floor

Questions 6 and 7 refer to the following information.

There were 3,342 students enrolled at a university. Of those students, 587 graduated in May. Over the summer, 32 students left the university, and 645 new students enrolled in the fall.

6. How many students were enrolled in the fall?

   (1) 2,697
   (2) 2,755
   (3) 3,310
   (4) 3,368
   (5) 3,987

7. Which number describes the change in the number of students enrolled between May and the following fall?

   (1) −32
   (2) −26
   (3) 26
   (4) 32
   (5) 58

Question 8 refers to the table below.

Melanie played a game and kept track of her score. The table shows her points earned for each round.

MELANIE'S POINTS SCORED	
ROUND	POINTS SCORED
1	8
2	−6
3	−4
4	3
5	4

8. What was Melanie's score at the end of Round 5?

   (1) 25
   (2) 13
   (3) 7
   (4) 5
   (5) 4

Algebraic Expressions and Variables

① Learn the Skill

A **variable** is a letter used to represent a number. Variables are used in algebraic expressions. An algebraic expression has numbers and variables possibly connected by an operation sign. A variable may change in value, which allows the expression itself to have different values. When you evaluate an algebraic expression, you substitute a number for the variable and solve. For example, if $b = 3$, then $b + 12 = 15$. If $b = -1$, then $b + 12 = 11$.

② Practice the Skill

Understanding how to use variables and how to simplify and evaluate algebraic expressions are important skills for success on the GED Mathematics Test. Read the example and strategies below. Then answer the question that follows.

A Order is important for division and subtraction. For example, "6 less than 3" is $3 - 6$, but "the difference between 6 and 3" is $6 - 3$.

B To simplify an expression, add like terms. Like terms have the same variable or variables raised to the same power. For example, $2x$ and $4x$ are like terms.

　　If an expression has parentheses, use the distributive property to simplify.

　　To evaluate an expression, substitute the given values for the variables, and then follow the order of operations.

WORDS	SYMBOLS
4 more than a number	$4 + x$
5 less than a number	$x - 5$
3 times a number	$3x$
A number times itself	x^2
The product of 8 and a number	$8x$
The product of 6 and x added to the difference between 5 and x	$6x + (5 - x)$
The quotient of 6 and x	$\frac{6}{x}$ or $6 \div x$
One-third of a number increased by 5	$\frac{1}{3}x + 5$

Simplify $4x(5x + 7) - 2x$

$(4x)(5x) + (4x)(7) - 2x$
$20x^2 + 28x - 2x$
$20x^2 + 26x$

☑ TEST-TAKING TIPS

Multiplication can be written in several ways. In algebraic expressions, a number next to a variable means multiplication. The expression $3y$ is the same as 3 times y. Parentheses and a dot (·) also indicate multiplication.

1. Gabe's current age is 3 times his sister's current age. Which expression represents Gabe's current age?

(1) $3x$

(2) $\dfrac{x}{3}$

(3) $x - 3$

(4) $x + 3$

(5) $3 \div x$

Directions: Choose the <u>one best answer</u> to each question.

2. The width of Kevin's yard is twice the width of his garage, increased by 10 feet. Which expression below describes the width of his yard if *g* represents the width of his garage?

   (1) 2*g*(10)

   (2) $\frac{2g}{10}$

   (3) 2*g* − 10

   (4) 2*g* + 10

   (5) *g* + 10

3. The number of employees that work in manufacturing is 500 less than 3 times the number of employees that work in shipping. Which expression represents the number of employees who work in manufacturing?

   (1) 3*s* + 500

   (2) 3*s*(500)

   (3) 3*s* − 500

   (4) $\frac{3s}{500}$

   (5) $\frac{500}{3s}$

4. Michael's score on a math quiz was 8 more than one-half of his score on his science quiz. Which expression below describes Michael's score on his math quiz?

   (1) $\frac{s}{2} + 8$

   (2) $\frac{s}{8} + 2$

   (3) $\frac{1}{2}s - 8$

   (4) $\frac{1}{2}(8) + s$

   (5) $\frac{s}{8} - \frac{1}{2}$

5. Julie left a tip on a restaurant bill that was one-sixth of the bill plus $2. If the bill was $48, how much was the tip?

   (1) $2
   (2) $4
   (3) $6
   (4) $8
   (5) $10

6. The cost of an adult ticket to the ballet is 2 times the cost of a child's ticket decreased by $4. If a child's ticket is $12, how much is an adult ticket?

   (1) $4
   (2) $8
   (3) $12
   (4) $20
   (5) $24

<u>Questions 7 and 8</u> refer to the figure below.

7. Which expression represents the perimeter of the rectangle?

$2w - 3$

   (1) 3*w* − 3
   (2) *w*(2*w* − 3)
   (3) 5*w* − 3
   (4) 4*w* − 6
   (5) 6*w* − 6

8. Which expression represents the area of the rectangle?

   (1) *w* + 2*w* − 3
   (2) *w*(2*w* − 3)
   (3) w^2
   (4) *w* + 2*w* − 3 + *w* + 2*w* − 3
   (5) Not enough information is given.

Equations

① Learn the Skill

An **equation** is a mathematical statement that shows two equal quantities. An equation shows an **expression** on each side of an equal sign. Notice that expressions do not include equal signs. An equation may or may not contain variables.

② Practice the Skill

To solve an equation, find the value of the variable that makes the statement true. To do this, isolate the variable on one side of the equation. Perform inverse operations to isolate the variable. Remember, addition and subtraction are inverse operations, as are multiplication and division. Read the examples and strategies below. Then answer the question that follows.

A Perform inverse operations on *both* sides of the equation. When performing an operation on one side of an equation, do the same to the other side. Perform the inverse operations for addition and subtraction first, and then for multiplication and division. When finished, substitute your solution for the variable into the equation to check your answer.

B Notes on Solving Equations:
- Some equations can be simplified before you solve. Combine like terms on either side of the equation.
- Some equations have variables on both sides. In this case, group all of the variables on one side.

Equations	**Expressions**
$4x + 8x = 36$ ⟶	$4x + 8x$
$3 = 6(x + 3) + 1$ ⟶	$6(x + 3) + 1$
$4 = 3 + 1$ ⟶	4

Solve an Equation	**Check**
$\frac{x}{-2} + 4 = 8$	$\frac{-8}{-2} + 4 = 8$
$\frac{x}{-2} + 4 - 4 = 8 - 4$	$4 + 4 = 8$
$\frac{x}{-2} = 4$	$8 = 8$
$(-2)\frac{x}{-2} = (-2)4$	
$x = -8$	

☑ TEST-TAKING TIPS

When a number is being multiplied by a variable, the number is called a *coefficient*. For example, the coefficient of $5x$ is *5*. If there is no number shown in front of the variable, the coefficient is *1*. For example, the coefficient of *y* is *1*.

1. Levi paid two bills. The cost of the two bills was $157. The second bill was $5 more than twice the amount of the first bill. Which of the following equations could be used to find the amount of the first bill?

   (1) $5 - 2x = 157$
   (2) $3x - 5 = 157$
   (3) $2x - 5 = 157$
   (4) $x - (2x + 5) = 157$
   (5) $x + (2x + 5) = 157$

Directions: Choose the <u>one best answer</u> to each question.

2. The sum of two consecutive integers is 15. Which equation could be used to find the first number?

   (1) $x + 2x = 15$

   (2) $2x + 1 = 15$

   (3) $x - 1 = 15$

   (4) $\frac{1}{2}x + 1 = 15$

   (5) $\frac{1}{x} - 1 = 15$

3. Caroline has twice as many science classes as literature classes. If she is taking 3 science and literature classes, which of the following equations could be used to find the number of literature classes she is taking?

   (1) $3x = 3$
   (2) $2x = 3$
   (3) $3x - 1 = 3$
   (4) $2x - 1 = 3$
   (5) $x = 3$

4. Stephanie's age is 3 years greater than half of her sister's current age. If her sister is 24 years old, what is Stephanie's age?

   (1) 12
   (2) 14
   (3) 15
   (4) 17
   (5) 21

5. The number of cellos in an orchestra is equal to 2 more than one-third of the number of violins. If there are 24 violins in the orchestra, how many cellos are there?

   (1) 5
   (2) 6
   (3) 8
   (4) 9
   (5) 10

6. Francisco has a total of twenty $5 and $1 bills in his wallet. The total value of the bills is $52. How many $5 bills does Francisco have in his wallet?

   (1) 8
   (2) 7
   (3) 6
   (4) 5
   (5) Not enough information is given.

7. Four times a number is four less than two times the number. What is the number?

   (1) −4
   (2) −2
   (3) 2
   (4) 4
   (5) Not enough information is given.

8. Julian collects rare political party convention pins. The number of Democratic Party pins he has is 14 less than 3 times the number of Republican Party pins he has. If he has 98 pins in all, how many Republican Party pins does he have?

   (1) 14
   (2) 28
   (3) 42
   (4) 70
   (5) 98

<u>Question 9</u> refers to the following figure.

The perimeter of the triangle below is 16.5 feet.

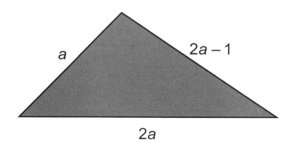

9. Which equation can be used to find the value of a?

   (1) $5a - 1 = 16.5$
   (2) $2a - 1 + a + 2a$
   (3) $a(2a - 1)(2a) = 16.5$
   (4) $4a - 1 = 16.5$
   (5) $2a - 1 + a + 2a + 16.5$

Exponents and Square Roots

① Learn the Skill

Exponents are used when a number, the **base**, is multiplied by itself many times. The exponent shows the number of times that the base is multiplied. When you raise a number to the second power, it is called squaring the number. To find the **square root** of a number, find the number that, when multiplied by itself, equals the given number. **Scientific notation** uses exponents and powers of 10 to write very small and very large numbers.

② Practice the Skill

A number or quantity raised to the first power equals itself. A number or quantity (except zero) raised to the zero power equals one. Terms only can be added and subtracted if they are like, meaning they must have the same variable raised to the same exponent. Read the examples and strategies below. Then answer the question that follows.

A To multiply terms with the same base, keep the base and add the exponents. Do the opposite for division. If the bases are not the same, simplify using the order of operations.

B To write a number shown in scientific notation as a number in expanded form, look at the power of 10. The exponent tells how many places to move the decimal point—right for positive, left for negative. To write a number in scientific notation, place the decimal point directly after the ones digit. Next, count the number of places you need to move. Then drop the zeros at the ends.

$$5^1 = 5 \qquad 5^0 = 1 \qquad 5^{-2} = \frac{1}{5^2} = \frac{1}{25}$$

$$2x^2 + 4x^2 + 1 = 6x^2 + 1 \qquad 4x^2 - x^2 = 3x^2$$

$$(3^2)(3^3) = (3)^{2+3} = 3^5 \qquad \frac{6^5}{6} = 6^{5-1} = 6^4$$

$$4.2 \times 10^7 = 42{,}000{,}000 \qquad 5{,}800{,}000 = 5.8 \times 10^6$$

$$3.7 \times 10^{-5} = 0.000037 \qquad 0.000052 = 5.2 \times 10^{-5}$$

$$4^2 = 16 \qquad 7^2 = 49 \qquad 11^2 = 121$$
$$\sqrt{16} = 4 \qquad \sqrt{49} = 7 \qquad \sqrt{121} = 11$$

☑ TEST-TAKING TIPS

You can use a calculator for exponents and square roots.
- square root of 45: 45 SHIFT x^2
- 45 squared: 45 x^2
- 45 to the 4th power: 45 x^y 4

1. The distance between the sun and Mercury is about 58,000,000 km. What is this distance written in scientific notation?

(1) 5.8×10^6
(2) 5.8×10^7
(3) 58×10^6
(4) 58×10^7
(5) 5.8×10^8

Directions: Choose the <u>one best answer</u> to each question.

2. Carlos completed 4^3 squats as part of his football workout. How many squats did he complete?

   (1) 8
   (2) 16
   (3) 36
   (4) 60
   (5) 64

3. There are 25,400,000 nanometers in an inch. What is this number written in scientific notation?

   (1) 2.54×10^6
   (2) 2.54×10^7
   (3) 2.54×10^8
   (4) 2.54×10^9
   (5) 2.54×10^{10}

4. The length of a square can be determined by finding the square root of its area. If a square has an area of 81 m^2, what is the length of the square?

   (1) 7.5 m
   (2) 8.0 m
   (3) 8.5 m
   (4) 9.0 m
   (5) 9.5 m

5. To determine the length of yarn needed for a project, Josie must solve $\frac{\sqrt{x}}{4}$ for $x = 64$. What is the solution?

   (1) 2
   (2) 3
   (3) 4
   (4) 5
   (5) 6

6. The width of a rectangle is 2^6, and the length is 2^5. What is the area of the rectangle?

   (1) 2^1
   (2) 2^{11}
   (3) 2^{30}
   (4) 4^1
   (5) 4^{11}

7. The width of a piece of blond hair is about 1.5×10^{-3} cm. What is the width of 2.0×10^5 blond hairs placed next to each other?

   (1) 3.5×10^8 cm
   (2) 3.0×10^{-2} cm
   (3) 3.0×10^8 cm
   (4) 3.0×10^2 cm
   (5) 5.0×10^{-8} cm

8. Which has the same value as $5^1 + 4^0$?

   (1) 9
   (2) 8
   (3) 6
   (4) 5
   (5) 4

9. Mark multiplied a number by itself. He found a product of 30. What is the number, rounded to the nearest tenth?

   (1) 4.5
   (2) 5.4
   (3) 5.5
   (4) 10.0
   (5) 15.0

10. A square has an area of 50 square feet. What is the perimeter of the square, rounded to the nearest foot?

   (1) 7 ft
   (2) 14 ft
   (3) 28 ft
   (4) 49 ft
   (5) 100 ft

Patterns and Functions

① Learn the Skill

A **mathematical pattern** is an arrangement of numbers and terms created by following a specific rule. You can identify the rule used to make a pattern and apply it to find other terms in the pattern. An algebraic rule is often called a **function**. A function contains x- and y-values. There is only one y-value for every x-value. Think of a function as a machine. For every x-value you put into the machine, only one y-value will come out.

② Practice the Skill

You will be asked to solve problems relating to patterns and functions on the GED Mathematics Test. Read the examples and strategies below. Then answer the question that follows.

Ⓐ To identify the rule used to make a pattern, study the sequence of numbers or terms. Ask yourself how each term is related to the next. In this example, the rule is "add 5."

Ⓑ A pattern also may be geometric. In this example, three triangles are added to the first figure to form the second figure. Five triangles are added to the second figure. Seven triangles are added to the third figure. The number of triangles to be added increases by two each time.

Ⓒ Functions are written as equations. They can show f(x) instead of y. Substitute each x-value (input) in the equation to find the value of f(x) (output). Remember, there is only one output for each input.

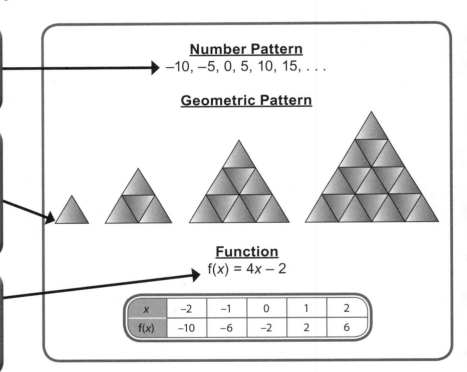

Number Pattern
−10, −5, 0, 5, 10, 15, . . .

Geometric Pattern

Function
f(x) = 4x − 2

x	−2	−1	0	1	2
f(x)	−10	−6	−2	2	6

✓ TEST-TAKING TIPS

You may need to test several rules to find the rule for a pattern. Look first for simple rules involving addition and subtraction. Then try multiplication and division. Some rules may involve more than one operation.

1. What is the value of f(x) if x = 4 in the function $f(x) = x^2 - 5$?

(1) −3
(2) −1
(3) 11
(4) 16
(5) 21

Directions: Choose the <u>one best answer</u> to each question.

<u>Questions 2 and 3</u> refer to the following sequence.

2, 4, 16, 256, 65,536, . . .

2. What is the rule for the pattern?

(1) multiply the previous term by 2
(2) add twice the previous term
(3) square the previous term
(4) divide by one-half
(5) multiply by 4

3. What is the next term in the sequence?

(1) 4,294,967,296
(2) 268,435,456
(3) 16,777,216
(4) 1,048,576
(5) 262,144

4. For the function $f(x) = \frac{x}{5}$, which of the following x-values has a whole number output?

(1) 19
(2) 21
(3) 22
(4) 25
(5) 26

5. The distance an airplane travels in t hours is given by the function $d = 230t$. How long does it take the airplane to travel 1,035 miles?

(1) 3.5 hours
(2) 4.5 hours
(3) 5.5 hours
(4) 6.5 hours
(5) 7.5 hours

6. Susan made a table to show the amount of sales tax due on typical purchase amounts. The function she used is $y = 0.08x$, where x is the cost of the purchase and y is the sales tax.

x	5	10	15	20	25
y	$0.40	$0.80	$1.20	$1.60	

How much sales tax does Susan owe if her purchases total $25?

(1) $3.20
(2) $2.80
(3) $2.40
(4) $2.00
(5) $1.80

7. What is the sixth term in the sequence below?

192, 96, 48, 24, . . .

(1) 1
(2) 3
(3) 6
(4) 12
(5) 15

8. Which shows the y-value that is missing from the table?

x	0	3	4	6	10
y	−1	14		29	49

(1) 3
(2) −3
(3) 19
(4) 20
(5) 29

9. If $f(x) = 2 - \frac{2}{3}x$, then what is x when $f(x) = 4$?

(1) −3

(2) $-\frac{4}{9}$

(3) $\frac{4}{9}$

(4) 3

(5) 9

Factoring

① Learn the Skill

Factors are numbers or expressions that are multiplied together to form a **product**. In the term $4y$, 4 and y are factors. A factor may have two terms, such as $(x + 5)$. You can multiply factors with two terms using the FOIL method, in which you multiply the *First, Inner, Outer,* and *Last* terms in that order. An expression or equation with more than one term also may be factored, or split into factors.

② Practice the Skill

A quadratic equation can be written in the form $ax^2 + bx + c = 0$, where a, b, and c are integers and a is not equal to zero. Knowing how to factor can help you solve for a missing value in a quadratic equation. Read the examples and strategies below. Then answer the question that follows.

A To factor, work backward.
1. List all of the possible factors for the third term.
2. Next, find the two factors of the third term that have a sum equal to the coefficient of the middle term.
3. Use the variable as the first term in each factor and the integers from step 3 as the second terms.
4. Use the FOIL method to check your answer.

B To solve a quadratic equation, rewrite the equation to set the quadratic expression equal to 0. Then factor and set each factor equal to 0. Then solve. Check both values by substituting them in the original equation.

The FOIL Method
Multiply $(x + 2)(x - 4)$

First $\quad x(x) = x^2$ $\qquad$ Outer $\quad x(-4) = -4x$
Inner $\quad 2(x) = 2x$ $\qquad$ Last $\quad 2(-4) = -8$

$$x^2 + (-4x) + 2x + (-8) = x^2 - 2x - 8$$

Factor $4x + 12$

$$\frac{4x + 12}{4} = \frac{4x}{4} + \frac{12}{4} = x + 3, \text{ so } 4x + 12 = 4(x + 3)$$

Factoring Quadratic Expressions
$$x^2 - 2x - 8$$
1. Factors of -8: $(1, -8), (-1, 8), (2, -4), (-2, 4)$
2. $-4 + 2 = -2$
3. $(x - 4)(x + 2)$
4. Check: $x^2 + 2x - 4x - 8 = x^2 - 2x - 8$

Solving Quadratic Equations
$x^2 + 4x = 12 \longrightarrow x^2 + 4x - 12 = 0 \longrightarrow (x - 2)(x + 6) = 0$
If $x - 2 = 0$ and $x + 6 = 0$, then $x = 2$ and $x = -6$.

✓ TEST-TAKING TIPS

Each term in an expression or equation belongs with the sign that precedes it. In the expression $x^2 - 2x - 8$, the first term is positive, while the next two terms are negative.

1. Which of the following are factors of $x^2 + 5x - 6$?

(1) $(x + 6)(x - 1)$
(2) $(x - 2)(x - 3)$
(3) $(x + 2)(x + 3)$
(4) $(x + 2)(x - 3)$
(5) $(x - 6)(x + 1)$

UNIT 3

Directions: Choose the <u>one best answer</u> to each question.

2. What is the product of $(x + 5)(x - 7)$?

   (1) $x^2 - 2x - 12$
   (2) $x^2 + 2x + 12$
   (3) $x^2 + 2x + 35$
   (4) $x^2 - 2x - 35$
   (5) $x^2 + 2x - 35$

3. Which of the following is equal to $(x - 3)(x - 3)$?

   (1) $x^2 - 6x + 9$
   (2) $x^2 + 6x - 9$
   (3) $x^2 + 6x + 9$
   (4) $x^2 - 9x - 6$
   (5) $x^2 - 9x + 6$

4. Which of the following is equal to $x^2 - 6x - 16$?

   (1) $(x + 4)(x - 4)$
   (2) $(x - 2)(x - 8)$
   (3) $(x - 2)(x + 8)$
   (4) $(x + 2)(x - 8)$
   (5) $(x - 4)(x - 4)$

5. The dimensions of a rectangle are $2x - 5$ and $-4x + 1$. Which expression represents the area of the rectangle?

$2x - 5$

$-4x + 1$

   (1) $-8x^2 + 22x - 5$
   (2) $8x^2 + 22x - 5$
   (3) $-8x^2 - 18x - 5$
   (4) $8x^2 - 18x - 5$
   (5) $-8x^2 + 22x + 5$

6. If $4x + 1$ is one factor of $4x^2 + 13x + 3$, which of the following is the other factor?

   (1) $x + 1$
   (2) $x + 3$
   (3) $x - 13$
   (4) $x + 12$
   (5) $x - 2$

7. Which shows the solutions for the following equation?

$$2x^2 + 18x + 36 = 0$$

   (1) -3 and 6
   (2) 3 and -6
   (3) 6 and 6
   (4) -6 and -6
   (5) -3 and -6

8. Miranda made a rectangular vegetable garden next to her house. She used the house as one side and fenced in the other three sides. She used 12 meters of fencing. The area of her garden is 32 square meters. To find a possible width of her garden, solve $w^2 - 12w = -32$.

   Which of the following is a possible width of her garden?

   (1) 1 m
   (2) 4 m
   (3) 6 m
   (4) 7 m
   (5) 10 m

9. Which expression has a product that has only two terms?

   (1) $(x + 7)(x - 1)$
   (2) $(x - 1)(x - 1)$
   (3) $(x - 7)(x + 7)$
   (4) $(x + 1)(x - 7)$
   (5) $(x - 7)(x - 7)$

Solving and Graphing Inequalities

① Learn the Skill

An **inequality** states that two algebraic expressions are not equal. Inequalities are written with less than and greater than symbols (<, >) as well as two additional symbols. The $\geq$ symbol means "is greater than or equal to" and the $\leq$ symbol means "is less than or equal to."

② Practice the Skill

A solution to an inequality can include an infinite amount of numbers. For example, solutions to $b < 5$ include $b = 4.5, 4, 3.99, 3, 2, 1, 0, -3, -10$, and so on. When each individual solution is plotted as a point on a number line, a solid line is formed, which represents the solution set. Read the examples and strategies below. Then answer the question that follows.

(A) Solve inequalities as you do equations. If you multiply or divide an inequality by a negative number, you must reverse the sign of the inequality. For example, if the inequality shown was $16 \leq -8x$, you would divide by -8 and reverse the sign, giving you $-2 \geq x$.

(B) For $x > 3$, every number to the right of 3 is in the solution set. Draw an open circle at 3 because 3 is *not* greater than 3 and therefore is not included in the solution set. Then draw a solid arrow to the right from 3.

For $x \leq 3$, each number to the left of 3 *as well as* 3 is included in the solution set. Draw a closed circle at 3 to show that 3 is included. Then draw a solid arrow pointing to the left from 3.

Examples of Inequalities

$x \geq 4$ ⟶ A number is greater than or equal to 4.

$2x + 7 < 15$ ⟶ Two times a number plus seven is less than 15.

Solving an Inequality

$$4 - 6(x - 3) \leq 2x + 6$$

Simplify $4 - 6(x - 3)$ $4 - 6x + 18 \leq 2x + 6$

$22 - 6x \leq 2x + 6$ ⟵ Add 6x to both sides.

$22 \leq 8x + 6$ ⟵ Subtract 6 from both sides.

$16 \leq 8x$ ⟵ Divide by 8.

$2 \leq x$, or $x \geq 2$

Graphing an Inequality

$x > 3$

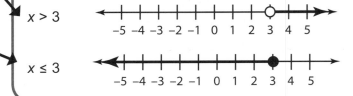

$x \leq 3$

✓ TEST-TAKING TIPS

When graphing an inequality with a variable on the right side of the inequality symbol, read the inequality backward. For example, think of $7 < y$ as "y is greater than 7."

1. Five times a number is less than or equal to two times the number plus nine. What is the solution to the inequality?

   (1) $x \geq 8$
   (2) $x \geq 9$
   (3) $x \leq 9$
   (4) $x \geq 3$
   (5) $x \leq 3$

Directions: Choose the <u>one best answer</u> to each question.

2. What is the solution to the inequality $x + 5 > 4$?

   (1) $x > 1$
   (2) $x < -1$
   (3) $x < 1$
   (4) $x > -1$
   (5) $x > x - 1$

3. What inequality is shown on the number line?

   (1) $x \leq 2$
   (2) $x \leq -2$
   (3) $x > 2$
   (4) $x < -2$
   (5) $x > -2$

4. The product of a number and 5, increased by 3, is less than or equal to 13. What is the inequality?

   (1) $5x + 2 \leq 13$
   (2) $5x \leq 13 + 3$
   (3) $5x + 3 < 13$
   (4) $5x + 3 \leq 13$
   (5) $x + 3 \leq 5x + 13$

5. The area of the following rectangle cannot be greater than 80 square centimeters. The length is 3 less than 3 times the width. Which inequality shows this relationship?

   w

   $3w - 3$

   (1) $80 \leq 2(w + 3w - 3)$
   (2) $80 \geq 2(w + 3w - 3)$
   (3) $80 \geq w(3w) - 3$
   (4) $80 \leq w(3w - 3)$
   (5) $80 \geq w(3w - 3)$

6. Kara has $15 and Brett has $22. Together, they have less than the amount needed to buy a pair of concert tickets. Which inequality describes their situation?

   (1) $37 < x$
   (2) $x + 15 < 22$
   (3) $x \leq 37$
   (4) $x + 22 \leq 15$
   (5) $x + 37 \leq 22$

7. A taxicab charges $2.00 as a base price and $0.50 for each mile. Josie needs to take a taxicab but only has $8. What is the greatest number of miles that Josie can ride in the cab?

   (1) 8
   (2) 9
   (3) 11
   (4) 12
   (5) 13

8. The sum of a number and 12 is less than or equal to 5 times the number plus 3. Which inequality represents this situation?

   (1) $x + 12 \geq 5(x - 3)$
   (2) $x + 12 \leq 5x + 3$
   (3) $x + 12 > 5x + 3$
   (4) $5x + 3 > x + 12$
   (5) $x + 3 \leq 15$

9. Which shows the solution for the inequality $8 - 3x > 2x - 2$?

   (1) $x > 2$
   (2) $x < 2$
   (3) $x > 6$
   (4) $x < 6$
   (5) $x > 8$

10. What is the solution to the inequality $-x - 4x > 30 - 3(x + 8)$?

   (1) $-6\frac{3}{4} > x$

   (2) $-3 < x$

   (3) $-6\frac{3}{4} < x$

   (4) $-3 > x$

   (5) $3 > x$

The Coordinate Grid

① Learn the Skill

A **coordinate grid** is a visual representation of points, or ordered pairs. An **ordered pair** is a pair of values: an *x*-value and a *y*-value. The *x*-value is always shown first. The grid is made by the intersection of a horizontal line (*x*-axis) and a vertical line (*y*-axis). The point where the number lines meet is called the **origin**, which is (0, 0). The grid is divided into four **quadrants**, or sections.

② Practice the Skill

The upper-right section of a grid is the first quadrant. Move counterclockwise to name the remaining quadrants. In an ordered pair, the first value (*x*-value) tells how many spaces to move (right for positive or left for negative). The *y*-value tells how many spaces to move (up for positive or down for negative). Examine the grid and strategies below. Then answer the question that follows.

A To draw a line segment on the coordinate grid, plot the given points. Then draw a line to connect them.

B Changes to figures can be shown on a coordinate grid. They include translations, reflections, rotations, and dilations. In a *translation*, a figure slides to a new position. In a *reflection*, a figure is flipped over a line such as an axis. In a *rotation*, a figure is turned about a point. In a *dilation*, a figure is proportionally resized.

The coordinate grid below shows points *A*, *B*, *C*, and *D*. The coordinates for point *A* are (−1, 0). Point *B* is located at (−1, −5).

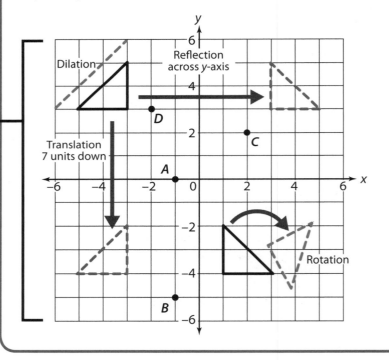

☑ TEST-TAKING TIPS

You will respond to certain questions on the GED Mathematics Test using a coordinate grid.

1. What are the coordinates of point *C*?

(1) (2, 2)
(2) (−2, 2)
(3) (2, −2)
(4) (−2, 3)
(5) (3, −2)

UNIT 3

③ Apply the Skill

Directions: Choose the <u>one best answer</u> to each question.

<u>Questions 2 through 4</u> refer to the following coordinate grid.

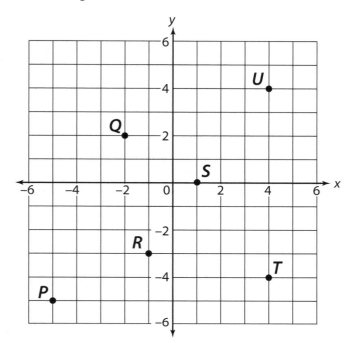

2. What are the coordinates of point *T*?

   (1) (5, −4)
   (2) (4, −4)
   (3) (4, −5)
   (4) (4, 4)
   (5) (−4, 4)

3. Which of the following ordered pairs describes the location of point *S*?

   (1) (1, −1)
   (2) (1, 0)
   (3) (−1, 0)
   (4) (0, 1)
   (5) (0, −1)

4. What are the coordinates of point *P*?

   (1) (−5, −5)
   (2) (−5, 5)
   (3) (5, −5)
   (4) (5, −6)
   (5) (6, −5)

<u>Questions 5 through 7</u> refer to the grid below.

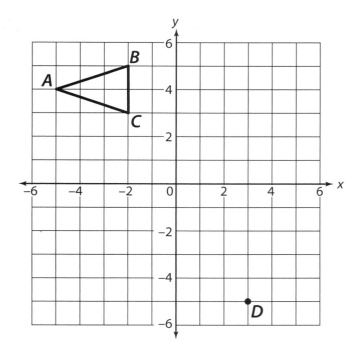

5. What are the coordinates of point *D*?

   (1) (3, 5)
   (2) (−3, −5)
   (3) (−5, 3)
   (4) (5, −3)
   (5) (3, −5)

6. If triangle *ABC* were reflected across the *y*-axis, what would be the new location of point *C*?

   (1) (1, 3)
   (2) (2, 5)
   (3) (2, 3)
   (4) (3, 3)
   (5) (2, 4)

7. If triangle *ABC* were translated 3 units down, what would be the new location of point *B*?

   (1) (−2, 2)
   (2) (−1, −5)
   (3) (−2, 0)
   (4) (−5, 1)
   (5) (−5, 5)

Graphing Equations

① Learn the Skill

Some equations have two variables. In this case, the value of one variable depends on the other. You can show the possible solutions for an equation with two variables on a graph. A **linear equation** is one that forms a straight line when graphed. All of the solutions of the equation lie on a line. To draw a line, you must find at least two points on the line and connect them.

② Practice the Skill

You will find questions relating to equations and graphing equations on the GED Mathematics Test. Read the examples and strategies below. Then answer the question that follows.

A Choose a value for *x*. Zero is an easy number with which to begin. Substitute the number for *x* and solve for *y*. This pair of values forms an ordered pair that lies on the graph of the line. Choose another value for *x* and solve for *y* to find another ordered pair. Plot and connect the two points to graph the line.

B Use this formula to find the distance between two points:

$$\text{distance between points} = \sqrt{(x_2 - x_1)^2 + (y_2 - y_1)^2}$$

To find the distance to the nearest tenth between points (0,−3) and (2, 5), substitute the coordinates into the formula. Solve.

$d = \sqrt{(2 - 0)^2 + (5 - (-3))^2}$
$ = \sqrt{2^2 + 8^2}$
$ = \sqrt{4 + 64}$
$ = \sqrt{68}$
$ \approx 8.2$

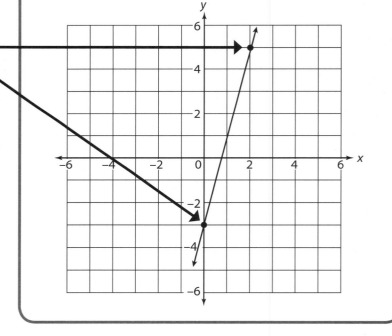

Graph $y = 4x - 3$

Let $x = 0$
$y = 4(0) - 3$
$y = 0 - 3$
$y = -3$
Plot (0,−3)

Let $x = 2$
$y = 4(2) - 3$
$y = 8 - 3$
$y = 5$
Plot (2, 5)

☑ TEST-TAKING TIPS

The expression $(-3)^2$ involves squaring −3, which equals 9; -3^2 involves squaring 3 and multiplying by −1, to equal −9.

1. Which ordered pair is a solution to $2x + y = 5$?

(1) (−1, 3)
(2) (3, −1)
(3) (0, −5)
(4) (−3, −4)
(5) (−2, 6)

UNIT 3

Question 6 refers to the following coordinate grid.

Directions: Choose the <u>one best answer</u> to each question.

2. Which of the following ordered pairs is a point on the line of the equation $x + 2y = 4$?

    (1) $(-2, 0)$
    (2) $(1, 3)$
    (3) $(0, 2)$
    (4) $(-3, -1)$
    (5) $(2, -4)$

3. Which ordered pair is a solution to $2x - y = 0$?

    (1) $(0, 0)$
    (2) $(1, -2)$
    (3) $(0, 2)$
    (4) $(-1, 2)$
    (5) $(2, -2)$

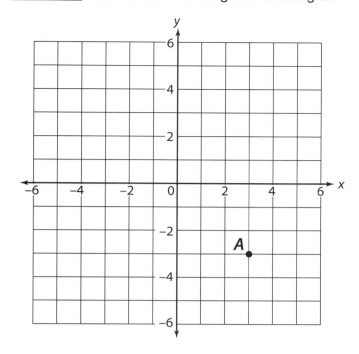

4. What is the missing x-value if $(x, 3)$ is a solution to $y = 2x + 2$?

    (1) -1

    (2) $-\dfrac{1}{2}$

    (3) 0

    (4) $\dfrac{1}{2}$

    (5) 1

6. Point A lies on a line of the equation $x + 2y = -3$. Which of the following is another point on this line?

    (1) $(0, -3)$
    (2) $(-1, 2)$
    (3) $(0, -2)$
    (4) $(-5, 1)$
    (5) $(4, -3)$

7. Two points are located at $(2, 5)$ and $(4, 3)$. What is the distance between the points to the nearest hundredth?

    (1) 1.41
    (2) 2.00
    (3) 2.45
    (4) 2.65
    (5) 2.83

5. A segment is drawn from the origin to $(-4, 3)$. What is the length of the segment?

    (1) 1.0
    (2) 2.6
    (3) 5.0
    (4) 7.0
    (5) 12.0

8. Marvin walks a straight line from $(-5, 2)$ to $(-3, 1)$ and stops. Then he walks a straight line from $(-3, 1)$ to $(-1, -4)$. What distance did Marvin walk?

    (1) 14.94
    (2) 9.04
    (3) 7.62
    (4) 6.00
    (5) 5.83

UNIT 3

Slope

① Learn the Skill

The **slope** is a number that measures the steepness of a line. Slope can be positive, negative, or zero. You can find the slope of a line by counting spaces on a graph or by using an algebraic formula. You also can use the slope-intercept form of a line to help you identify the equation of a line.

② Practice the Skill

The slope of a line is constant, meaning that the line always climbs or falls at the same rate. A line that rises from left to right has a positive slope. A line that falls from left to right has a negative slope. A horizontal line has a slope of zero, and a vertical line has an undefined slope, or no slope. If two lines are parallel, they have the same slope. Examine the grids and strategies below. Then answer the question that follows.

A Use two points to find a slope. Start at the lower point. How many units must you climb to reach the other point? This is the *rise*, or numerator. How many units must you move left or right to reach the point? This is the *run*, or denominator. If you move left, the value is negative. There is also an algebraic formula you can use to find slope.

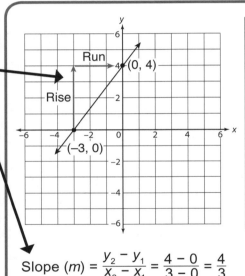

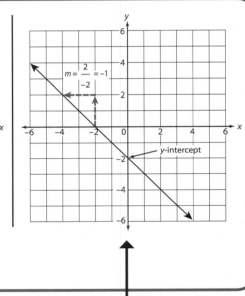

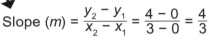

Slope $(m) = \dfrac{y_2 - y_1}{x_2 - x_1} = \dfrac{4 - 0}{3 - 0} = \dfrac{4}{3}$

B To find the equation of a line, find the *y*-intercept (where the line crosses the *y*-axis). The line crosses the *y*-axis at −2. Next, find the slope. The slope of this line is −1. Substitute the values of *m* and *b* into the equation.
$$y = mx + b$$
$$y = -1x + (-2)$$
$$y = -x - 2$$

☑ TEST-TAKING TIPS

Notice that the slopes of $\frac{-1}{2}$ and $\frac{1}{-2}$ have the same value. Both show a negative slope. However, $\frac{-1}{-2}$ actually shows a positive slope, because a negative divided by a negative equals a positive.

1. What is the slope of a line that passes through (−1, 3) and (1, 4)?

   (1) $-\dfrac{1}{2}$

   (2) 0

   (3) $\dfrac{1}{2}$

   (4) $\dfrac{2}{3}$

   (5) 1

Directions: Choose the <u>one best answer</u> to each question.

<u>Questions 2 and 3</u> refer to the grid below.

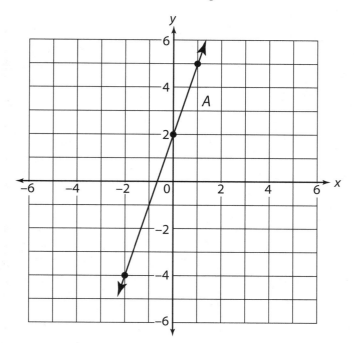

2. The following points lie on line *A*: (−2, −4), (0, 2), and (1, 5). What is the slope of line *A*?

    (1) 3
    (2) 2
    (3) 1
    (4) −2
    (5) −3

3. What is the equation of line *A*?

    (1) $y = \frac{1}{2}x + 3$

    (2) $y = \frac{1}{3}x + 2$

    (3) $y = 2x + 3$
    (4) $y = 3x + 2$
    (5) $y = 3x + 3$

4. A ramp was built to allow wheelchair access to a front door. The ramp rises 2 feet, as shown in the diagram below.

    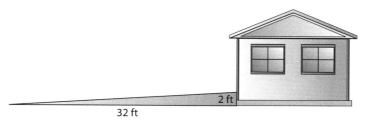

    What is the slope of the ramp?

    (1) $\frac{1}{32}$

    (2) $\frac{1}{18}$

    (3) $\frac{1}{16}$

    (4) $\frac{1}{8}$

    (5) $\frac{1}{4}$

5. A linear function is represented by f(*x*) = 2. What is the slope of the line?

    (1) −2
    (2) −1
    (3) 0
    (4) 1
    (5) 2

6. Which equation shows a line parallel to 4 − *y* = 2*x*?

    (1) 2 + *y* = 2*x*

    (2) $y - 2 = \frac{1}{2}x$

    (3) $2 + \frac{1}{2}y = 2x$

    (4) −*y* = 2 − 2*x*

    (5) *y* = −2*x* + 2

Unit 3 Review

On the GED Mathematics Test you will be asked to write your answers in different ways. Below are three ways to write your answers for this Unit Review.

Horizontal-response format

①②●④⑤

To record your answers, fill in the numbered circle that corresponds to the answer you select for each question in the Unit Review. Do not rest your pencil on the answer area while considering your answer. Make no stray or unnecessary marks. If you change an answer, erase your first mark completely. Mark only one answer space for each question; multiple answers will be scored as incorrect.

Alternate-response format

To record your answers for an alternate format question
- Begin in any column that will allow your answer to be entered;
- Write your answer in the boxes in the top row;
- In the column beneath a fraction bar or decimal point (if any) and each number in your answer, fill in the bubble representing that character;
- Leave blank any unused column.

Coordinate grid format

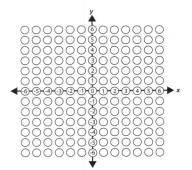

Points to consider when recording an answer on the coordinate grid:
- To record an answer, you must have an "x" value and a "y" value.
- No answer will have a value that is a fraction or a decimal.
- Mark only the one circle that represents your answer.

Directions: Choose the one best answer to each question.

1. A painter charges $20 per hour for herself and $15 per hour for her assistant. In painting a living room, the assistant worked 5 hours more than the painter. The total charge for labor was $355.

   Let h be the number of hours that the painter worked. Which of the following equations can be used to find h?

   (1) $20h + 15(h + 5) = 355$
   (2) $20(h + 5) + 15h = 355$
   (3) $20h + 15(h − 5) = 355$
   (4) $20h − 15(h + 5) = 355$
   (5) $20h − 15(h − 5) = 355$

   ①②③④⑤

2. If $x^2 = 36$, then $2(x + 5)$ could equal which of the following numbers?

   (1) 6
   (2) 11
   (3) 12
   (4) 22
   (5) 28

   ①②③④⑤

3. Sara has $1,244 in her checking account. She deposits a check for $287 and withdraws $50 cash. What is her new balance?

   (1) $1,294
   (2) $1,394
   (3) $1,481
   (4) $1,531
   (5) $1,581

   ①②③④⑤

4. What is the next term in the sequence?

3, 2.5, 2, 1.5, 1, 0.5, 0, . . .

(1) −1
(2) −0.5
(3) 0
(4) 1
(5) 1.5

①②③④⑤

5. The number of men acting in a theater production is five more than half the number of women. Which of the following expressions describes the number of men in the production?

(1) $2w + 5$

(2) $\frac{1}{2}w + 5$

(3) $2w - 5$

(4) $\frac{1}{2}w - 5$

(5) $-\frac{1}{2}w + 5$

①②③④⑤

6. If $3x + 0.15 = 1.29$, what is the value of x?

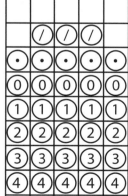

Questions 7 and 8 refer to the following grid below.

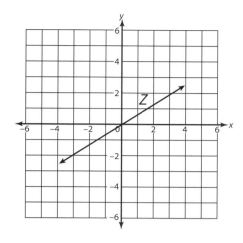

7. What is the slope of line Z?

(1) $-\frac{2}{3}$

(2) $-\frac{1}{2}$

(3) 1

(4) $\frac{1}{2}$

(5) $\frac{2}{3}$

①②③④⑤

8. What is the equation of line Z in slope-intercept form?

(1) $y = -\frac{2}{3}x + 1$

(2) $y = \frac{1}{2}x$

(3) $y = \frac{2}{3}x$

(4) $y = \frac{3}{2}x$

(5) $y = x + \frac{2}{3}$

①②③④⑤

9. The number of people voting in an election who were over 25 years old was 56 less than twice the number of those voting who were under 25 years old. Which expression represents the number of people who were over 25 years old who voted in the election?

(1) $56x - 25$
(2) $2x - 56$
(3) $x + 56$
(4) $56x + 25$
(5) $2x + 56$

①②③④⑤

10. Ellie has a pass that allows her to drive through tolls without stopping to pay. The amount of the toll is automatically charged to her credit card. She pays a fee of $5 per month for this service. Each toll she pays is $1.25. She budgets $65 a month for her total toll bill. What is the maximum number of tolls she can pass through each month and still stay within her budget?

(1) 12
(2) 24
(3) 36
(4) 48
(5) 60

①②③④⑤

11. The Earth is 149,600,000 kilometers from the sun. What is this distance written in scientific notation?

(1) 1.496×10^{-7} km
(2) 1.496×10^{-8} km
(3) 1.496×10^{8} km
(4) 1.496×10^{9} km
(5) 1.496×10^{10} km

①②③④⑤

Questions 12 and 13 refer to the following grid.

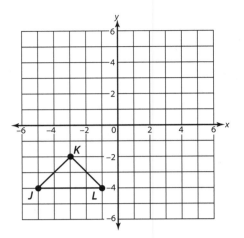

12. If triangle *JKL* is reflected across the *y*-axis, then what is the new location of point *K*?

Mark your answer on the coordinate plane grid.

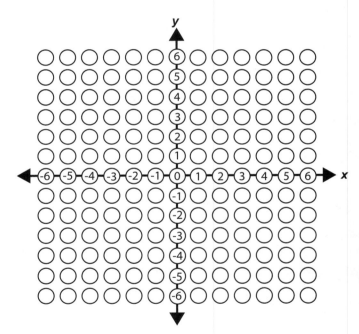

13. What is the slope of side *JL* in triangle *JKL*?

(1) −2
(2) −1
(3) 0
(4) 1
(5) 2

①②③④⑤

14. The product of two consecutive negative integers is 19 more than their sum. What is the greater integer?

(1) 0
(2) −1
(3) −2
(4) −3
(5) −4

①②③④⑤

15. Line *H* passes through points (−5, 4) and (0, 1). What is the distance between these two points to the nearest hundredth unit?

```
┌───┬───┬───┬───┬───┐
│   │   │   │   │   │
├───┼───┼───┼───┼───┤
│   │ / │ / │ / │   │
├───┼───┼───┼───┼───┤
│ • │ • │ • │ • │ • │
├───┼───┼───┼───┼───┤
│ 0 │ 0 │ 0 │ 0 │ 0 │
│ 1 │ 1 │ 1 │ 1 │ 1 │
│ 2 │ 2 │ 2 │ 2 │ 2 │
│ 3 │ 3 │ 3 │ 3 │ 3 │
│ 4 │ 4 │ 4 │ 4 │ 4 │
│ 5 │ 5 │ 5 │ 5 │ 5 │
│ 6 │ 6 │ 6 │ 6 │ 6 │
│ 7 │ 7 │ 7 │ 7 │ 7 │
│ 8 │ 8 │ 8 │ 8 │ 8 │
│ 9 │ 9 │ 9 │ 9 │ 9 │
└───┴───┴───┴───┴───┘
```

16. Each day for three days, Emmit withdrew $64 from his account. Which number shows the change in his account after the three days?

(1) −$192
(2) −$128
(3) −$64
(4) $128
(5) $192

①②③④⑤

17. The function $y = \frac{3}{4}x$ was used to create the following table. Which number is missing from the table?

x	−2	−1	0	1	2
y	$-\frac{3}{2}$	$-\frac{3}{4}$	0	$\frac{3}{4}$	

(1) $-\frac{1}{4}$

(2) $-\frac{1}{2}$

(3) 1

(4) $\frac{3}{2}$

(5) $\frac{5}{2}$

①②③④⑤

18. A skier takes a chairlift 786 feet up the side of a mountain. He then skis down 137 feet and catches a different chairlift 542 feet up the mountain. What is his position when he gets off the chairlift relative to where he began on the first chairlift?

(1) −1,191 feet
(2) −649 feet
(3) +679 feet
(4) +1,191 feet
(5) +1,465 feet

①②③④⑤

19. Which of the following inequalities is shown on the number line?

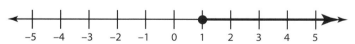

(1) $x \geq 1$
(2) $x \leq 1$
(3) $x < -1$
(4) $x > -1$
(5) $x > 1$

①②③④⑤

UNIT 3

20. A summer family camp costs $230 for adults. The cost for a child is $30 less than one-half the cost for adults. What is the cost for 3 children?

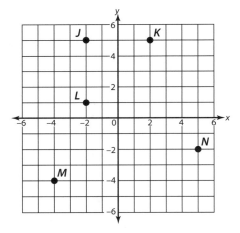

Questions 21 and 22 refer to the following grid.

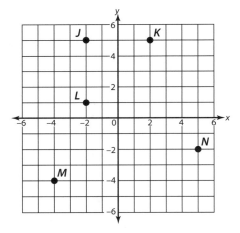

21. Points J, K, and L mark the corners of a rectangle. What is the location of the fourth corner needed to complete the rectangle?

Mark your answer on the coordinate plane grid.

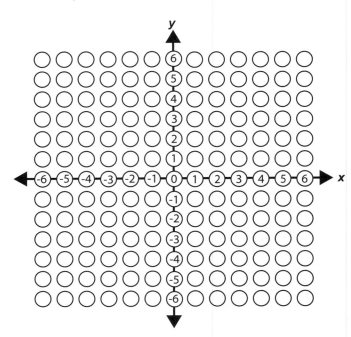

22. Which of the following is the equation of a line that passes through points L and K?

(1) $y = \frac{1}{3}x$

(2) $y = 3x$
(3) $y = x + 3$
(4) $y = x - 3$
(5) $y = 3x + 3$

①②③④⑤

23. The equation $h = -16t^2 - 48t + 160$ represents the height h of a ball above ground at time t seconds after being dropped. How many seconds does it take the ball to reach the ground?

(1) 1
(2) 2
(3) 3
(4) 4
(5) 5

①②③④⑤

24. The weight of a mother elephant is 200 kg more than 4 times the weight of her newborn calf. Which of the following expressions represents the mother elephant's weight?

(1) $4c + 200$
(2) $200 - 4c$
(3) $4(c - 200)$
(4) $4c - 200$
(5) Not enough information is given.

①②③④⑤

25. Keenan purchased solar lights for his front walkway. The total amount he paid for 8 lights was $73.36, including $4.16 tax. What was the cost per light before tax?

(1) $7.92
(2) $8.18
(3) $8.65
(4) $9.17
(5) $9.69

①②③④⑤

26. The number of students at a large university can be written as 8^4. How many students are at the university?

(1) 512
(2) 4,096
(3) 10,024
(4) 32,028
(5) 32,768

①②③④⑤

27. A ticket to a concert costs $32. Concert T-shirts are $12. Edward takes $80 to the concert. He has to buy his ticket, and he also wants to buy T-shirts for his family. Which inequality represents the possible number of T-shirts he can buy?

(1) $80 - 32 \leq x$
(2) $32 \geq 80 - x$
(3) $32x - 12 < 80$
(4) $12x + 32 \leq 80$
(5) $12x + 80 \leq 32$

①②③④⑤

Questions 28 and 29 refer to the following grid.

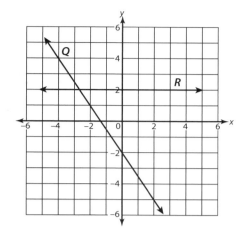

28. Which of the following is the equation of line R?

(1) $x = 2$
(2) $y = 2$
(3) $y = x + 2$
(4) $x = y + 2$
(5) $y = 2x$

①②③④⑤

29. What is the slope of line Q?

(1) $-\dfrac{3}{2}$
(2) $-\dfrac{1}{2}$
(3) $\dfrac{1}{2}$
(4) $\dfrac{3}{2}$
(5) 1

①②③④⑤

30. What is the distance between point $(-4, 2)$ and point $(5, 2)$?

(1) 5
(2) 6
(3) 7
(4) 8
(5) 9

①②③④⑤

UNIT 3

Unit 4

HUONG McDONIEL

Huong McDoniel's journey took her from Vietnam to the Philippines and Guam and on to Arizona and to her life today as a college faculty member in New Mexico.

Huong McDoniel left behind her books, but not her love of learning. As a teenager during the fall of Saigon, Huong McDoniel packed two suitcases: one with clothes, the other with science and math books. Before climbing to the top of the platform of the U.S. Embassy in Saigon, where helicopters were evacuating refugees, McDoniel was forced to leave her textbooks behind.

Flown to an aircraft carrier, McDoniel and 6,000 others then crowded onto a ship equipped for only 600 crew members. After living in refugee camps in the Philippines and Guam, McDoniel moved to Tucson, Arizona. At the time, she didn't yet speak English and first worked as a dishwasher. Her dream of becoming a teacher seemed as distant as her early life in Vietnam.

But then McDoniel moved to Albuquerque, New Mexico, where in 1989 she decided to pursue a GED certificate while raising three children and also learning English. With help from tutors at Central New Mexico (CNM) Community College, McDoniel passed the GED Tests on her first try. As she recalls,

> **❝ The more you learn, the more your dream begins to expand. Then you start to believe you can do more. Slowly, I started thinking maybe I could become a teacher. ❞**

McDoniel continued her studies at CNM, earning an associate's degree in liberal arts and herself becoming a tutor. She then enrolled at the University of New Mexico, from which she earned a degree in math and a master's degree in education. Today, McDoniel works as a full-time faculty member at CNM and speaks to GED graduates about their achievements.

BIO BLAST: Huong McDoniel

- Born in Vietnam and airlifted from Saigon during Operation Frequent Wind in 1975
- Married to Doug McDoniel, faculty member at Central New Mexico Community College
- Raised three children while earning her associate's degree and working as a college tutor
- Achieved her dream of becoming a teacher

Geometry

Unit 4: Geometry

Look around your home, office, or town. Chances are, wherever you look—furniture, rooms, buildings, neighborhoods, or cities—you'll see a variety of geometric figures. Geometry enables us to solve many everyday problems using points, lines, and angles, as well as figures such as squares, rectangles, triangles, circles, and solids.

Geometry similarly appears frequently on the GED Mathematics Test, accounting for between 20 and 30 percent of all questions. Among other concepts, in Unit 4 you will study lines and angles, a variety of regular and irregular figures, and circles and solids, all of which will help you prepare for the GED Mathematics Test.

Table of Contents

Lines and Angles

① Learn the Skill

A **line** continues forever in both directions. A **ray** is a part of a line. It has one **endpoint** and extends in the other direction without end. The first figure contains $\overrightarrow{AB}$ and $\overrightarrow{AC}$. The arrow above the letters means "ray." Two rays that share the same endpoint form an **angle**. The shared endpoint is called the **vertex**. In this figure, the vertex is point A. The angle in the first figure can be named ∠BAC, ∠CAB, ∠A, or ∠1.

Two angles that have the same measure are **congruent**. If the sum of the measures of the angles is 90°, the angles are **complementary**. The two angles shown directly to the right are complementary. If the sum of the measures of two angles is a straight angle, or 180°, the angles are **supplementary**.

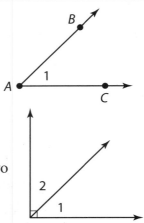

② Practice the Skill

By mastering the skills related to lines and angles, you will improve your study and test-taking skills, especially as they relate to the GED Mathematics Test. Read the example and strategies below. Then answer the question that follows.

A Angles that share a common side and have the same vertex are **adjacent angles**. In this figure, ∠1 and ∠2 are adjacent and supplementary. Angles formed by intersecting lines and that are not adjacent are called **vertical angles**. Vertical angles are congruent. In this figure, ∠5 and ∠7 are vertical angles. Which other pairs of angles are vertical angles?

B In this figure, ∠3, ∠4, ∠5, and ∠6 are **interior angles**. Angles 1, 2, 7, and 8 are **exterior angles**. Angles 3 and 5 are **alternate interior angles**. They are congruent. The same is true of ∠4 and ∠6. There are two pairs of **alternate exterior angles**: ∠1, ∠7 and ∠2, ∠8. These pairs of angles also have the same measure. Corresponding angles are also congruent. Angles 2 and 6 are corresponding because they are in the same relative location.

The figure below shows two parallel lines intersected by $\overleftrightarrow{JK}$. This line is called a transversal (a line that intersects parallel lines).

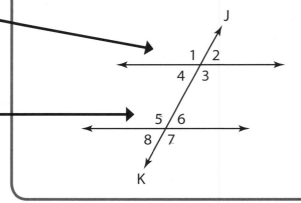

☑ TEST-TAKING TIPS

In the example, ∠1 and ∠5 are congruent because they are corresponding, and ∠6 and ∠5 are supplementary. Find 180 − 115 to solve the problem.

1. If ∠1 is 115°, what is the measure of ∠6?

   (1) 65°
   (2) 85°
   (3) 115°
   (4) 180°
   (5) Not enough information is given.

<u>Directions</u>: Choose the <u>one best answer</u> to each question.

<u>Questions 2 and 3</u> refer to the following text and figure.

Two lines intersect as shown in the diagram below.

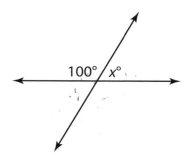

2. What is the value of *x*?

   (1) 80
   (2) 90
   (3) 100
   (4) 180
   (5) Not enough information is given.

3. What is the measure of the angle that is vertical to the angle measuring 100°?

   (1) 80°
   (2) 90°
   (3) 100°
   (4) 180°
   (5) Not enough information is given.

4. Two angles are complementary. If one angle has a measure of 25°, then what is the measure of the other angle?

   (1) 25°
   (2) 65°
   (3) 75°
   (4) 155°
   (5) 165°

<u>Questions 5 and 6</u> refer to the following text and figure.

This figure shows two parallel lines that are transversed, or crossed, by a third line. Remember that parallel lines are equidistant and never intersect.

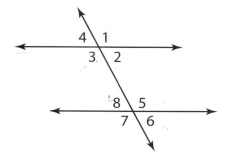

5. Which two angles have the same measure?

   (1) ∠1 and ∠4
   (2) ∠2 and ∠5
   (3) ∠7 and ∠8
   (4) ∠1 and ∠3
   (5) ∠1 and ∠6

6. Which three angles have the same measure?

   (1) ∠1, ∠2, and ∠3
   (2) ∠2, ∠5, and ∠6
   (3) ∠1, ∠3, and ∠7
   (4) ∠5, ∠7, and ∠8
   (5) ∠6, ∠8, and ∠3

<u>Question 7</u> refers to the following text and figure.

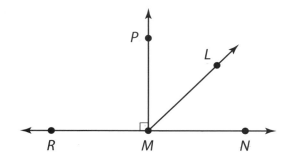

7. If ∠*PMR* is a right angle and ∠*LMN* is 35°, what is the sum of the measures of ∠*LMP* and ∠*NML*?

   (1) 35°
   (2) 55°
   (3) 90°
   (4) 125°
   (5) 180°

Triangles and Quadrilaterals

① Learn the Skill

A **triangle** is a closed three-sided figure with three angles. The sum of the three interior angles of any triangle is always 180°. To name a triangle, use the symbol △ followed by the names of the vertices in any clockwise or counterclockwise order. For example, two names for the triangle to the right include △CBA and △CAB.

A **quadrilateral** is a closed four-sided figure with four angles. The sum of the four interior angles of any quadrilateral is always 360°. The sides of a quadrilateral may or may not be parallel. This quadrilateral is a trapezoid, which has one pair of parallel sides. Other quadrilaterals are the parallelogram, rectangle, rhombus, and square.

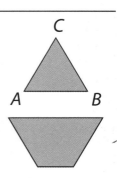

② Practice the Skill

By mastering your understanding of triangles and quadrilaterals, you will improve your study and test-taking skills, especially as they relate to the GED Mathematics Test. Read the example and strategies below. Then answer the question that follows.

Ⓐ Triangles can be classified by their largest angle: *right* (90°), *acute* (less than 90°), or *obtuse* (greater than 90°). They also can be classified by their sides:
   equilateral: three congruent sides
   isosceles: at least two congruent sides
   scalene: no congruent sides
This is a scalene triangle. No two sides are the same length, which means that no two angles have the same measure.

Ⓑ In a rectangle and square, all angles are right angles and are congruent. In a parallelogram and rhombus, opposite angles and opposite sides are congruent. You can use this information to find the measure of an unknown angle. In this parallelogram, ∠D and ∠B are congruent, and ∠A and ∠C are congruent.

The first figure is a triangle and the second figure is a parallelogram.

☑ **TEST-TAKING TIPS**

If you know the measure of any two angles of a triangle, you can determine the measure of the third angle by subtracting the sum of the two known angles from 180°. For example, in △XYZ, ∠Z measures 125°.

1. What is the measure of ∠A of the parallelogram?

   (1) 55°
   (2) 110°
   (3) 125°
   (4) 180°
   (5) 250°

Directions: Choose the <u>one best answer</u> to each question.

<u>Questions 2 and 3</u> refer to the following text and figure.

Supplementary angles are two angles whose measures add to 180°. Use this information and the figure below to answer the questions.

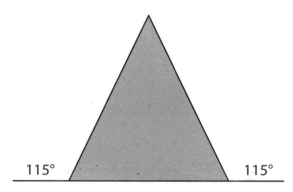

2. What is the measure of the smallest acute angle within the triangle?

   (1) 50°
   (2) 65°
   (3) 90°
   (4) 115°
   (5) 130°

3. Which statement best describes this isosceles triangle?

   (1) All angles are congruent.
   (2) All sides are congruent.
   (3) One angle is obtuse.
   (4) Two sides are congruent.
   (5) All sides and angles are congruent.

4. In a certain right triangle, the measure of one acute angle is twice the measure of the other acute angle. What is the measure of the smaller angle?

   (1) 10°
   (2) 25°
   (3) 30°
   (4) 55°
   (5) 60°

<u>Questions 5 and 6</u> refer to the following text and figure.

Quadrilateral *RSUV* is a parallelogram. Quadrilateral *RSTV* is a trapezoid.

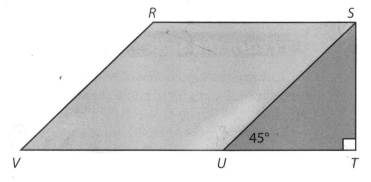

5. What is the measure of ∠*R*?

   (1) 45°
   (2) 90°
   (3) 135°
   (4) 145°
   (5) Not enough information is given.

6. What is the measure of ∠*RST*?

   (1) 22.5°
   (2) 45°
   (3) 65°
   (4) 90°
   (5) Not enough information is given

<u>Question 7</u> refers to the following text and figure.

Angle *H* in the rhombus shown is 35°.

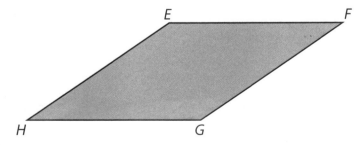

7. What is the measure of each obtuse angle in the rhombus?

   (1) 35°
   (2) 45°
   (3) 90°
   (4) 135°
   (5) 145°

Congruent and Similar Figures

① Learn the Skill

Congruent figures have the same shape and size. They have corresponding angles of equal measure and corresponding sides of equal length. The figures at the right are congruent figures.

Similar figures have the same shape and congruent angles, but have proportional sides. The similar triangles to the right have corresponding angles of equal measure but do not have corresponding sides of equal measure. If the base of the larger triangle is three times greater than the base of the other, the same relationship exists for the heights.

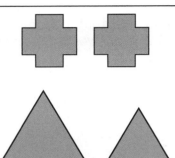

② Practice the Skill

If you know that two figures are similar, then you know that the corresponding angles are congruent and the sides are proportional. Therefore, if you know that two figures have congruent corresponding angles and proportional sides, then you know that the figures are similar. Read the example and strategies below. Then answer the question that follows.

A When angles or line segments of two figures correspond, they are in the same position. Angle C of △ABC corresponds to ∠T of △RST. Angle B corresponds to ∠S. Similarly, $\overline{AC}$ corresponds to $\overline{RT}$, and $\overline{BC}$ corresponds to $\overline{ST}$.

B The symbol ≅ means "is congruent to." The symbol ~ means "is similar to." When saying that two figures are congruent or similar, name corresponding parts in the same order. For example, △BCA ~ △STR.

Triangles *ABC* and *RST* are similar.

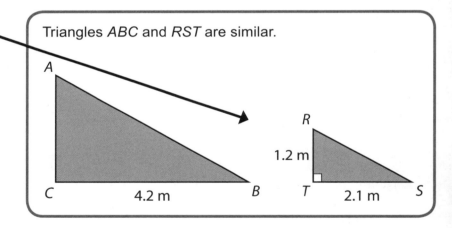

☑ TEST-TAKING TIPS

In question 1, we are told that the triangles are similar. A proportion can be written to solve for the missing length. A sample proportion for the figures is $\dfrac{x}{4.2} = \dfrac{1.2}{2.1}$.

1. What is the length of $\overline{AC}$?

(1) 1.2 m
(2) 2.1 m
(3) 2.4 m
(4) 3.2 m
(5) 4.2 m

③ Apply the Skill

Directions: Choose the <u>one best answer</u> to each question.

Question 2 refers to the following figures.

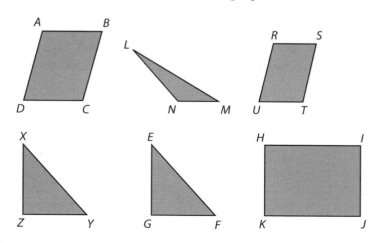

2. Which statement appears to be true?

   (1) parallelogram *ABCD* ~ rectangle *HIJK*
   (2) triangle *LMN* ~ △*EFG*
   (3) parallelogram *ABCD* ≅ rectangle *HIJK*
   (4) triangle *XYZ* ≅ △*EFG*
   (5) parallelogram *RSTU* ~ rectangle *HIJK*

Question 3 refers to the following text and figures.

Triangles 1 and 2 shown below are congruent. The lengths of two sides of triangle 1 are given.

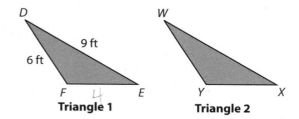

3. If the perimeter of triangle 1 is 19 ft, what is the length of $\overline{XY}$?

   (1) 4 ft
   (2) 6 ft
   (3) 9 ft
   (4) 19 ft
   (5) Not enough information is given.

Questions 4 and 5 refer to the following text and figures.

Triangle *ABC* and triangle *FGH* are similar figures.

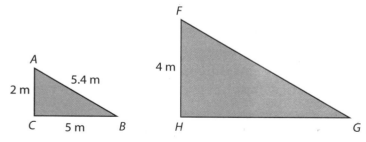

4. What is the length of $\overline{FG}$?

   (1) 5.4 m
   (2) 10.0 m
   (3) 10.8 m
   (4) 12.4 m
   (5) Not enough information is given.

5. What is the perimeter of △*FGH*?

   (1) 12.4 m
   (2) 14.8 m
   (3) 24.0 m
   (4) 24.8 m
   (5) Not enough information is given.

Question 6 refers to the following text and figures.

Figure *MNOP* and figure *XYZW* are congruent.

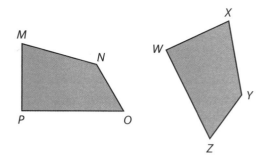

6. Which line segment of quadrilateral *XYZW* is congruent to $\overline{MN}$?

   (1) $\overline{WX}$
   (2) $\overline{XY}$
   (3) $\overline{YZ}$
   (4) $\overline{ZW}$
   (5) Not enough information is given.

Indirect Measurement and Proportion

① Learn the Skill

Indirect measurement involves the use of proportions and corresponding parts of similar figures to find a measurement that you cannot find directly. **Scale drawings**, or drawings that represent an actual object, can use indirect measurement.

② Practice the Skill

You will use indirect measurement to solve measurement and geometry problems on the GED Mathematics Test. Read the example and strategies below. Then answer the question that follows.

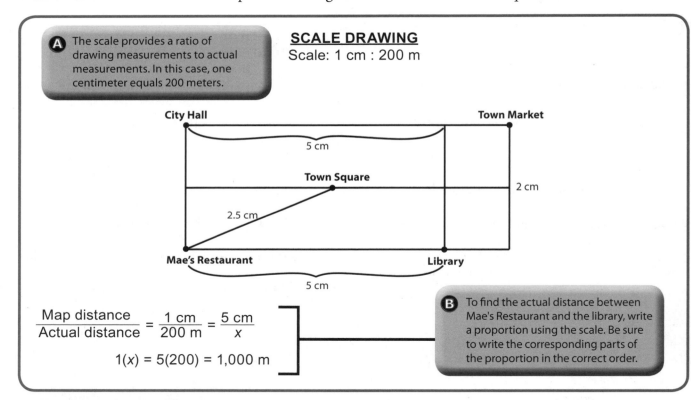

A The scale provides a ratio of drawing measurements to actual measurements. In this case, one centimeter equals 200 meters.

SCALE DRAWING
Scale: 1 cm : 200 m

City Hall   Town Market
5 cm
Town Square   2 cm
2.5 cm
Mae's Restaurant   Library
5 cm

$$\frac{\text{Map distance}}{\text{Actual distance}} = \frac{1 \text{ cm}}{200 \text{ m}} = \frac{5 \text{ cm}}{x}$$

$$1(x) = 5(200) = 1{,}000 \text{ m}$$

B To find the actual distance between Mae's Restaurant and the library, write a proportion using the scale. Be sure to write the corresponding parts of the proportion in the correct order.

✓ **TEST-TAKING TIPS**

Pay close attention to the units involved in a proportion. For example, a question posed about the map above could be "How many kilometers from City Hall is Town Market?" To answer this question correctly, you must convert meters to kilometers.

1. Two cities are 5 inches apart on a map. The map scale is 1 in. : 2.5 miles. What is the actual distance in miles between the two cities?

(1) 0.5
(2) 5
(3) 10
(4) 12.5
(5) 15

③ Apply the Skill 🖩

Directions: Choose the <u>one best answer</u> to each question.

2. A tree 14 feet tall casts a shadow that is 2.5 feet long. At the same time of day, a person casts a shadow that is 1 foot long. Which proportion can be solved to find the height of the person?

(1) $\frac{x}{1} = \frac{2.5}{14}$

(2) $\frac{1}{x} = \frac{2.5}{14}$

(3) $\frac{14}{1} = \frac{2.5}{x}$

(4) $\frac{1}{2.5} = \frac{14}{x}$

(5) Not enough information is given.

3. Erika drove from Plymouth to Manchester and back again. On a map, these two cities are 2.5 cm apart. If the map scale is 1 cm : 6 km, how many kilometers did she drive?

(1) 2.4
(2) 8.5
(3) 15
(4) 22.5
(5) 30

4. A map scale is 2 in. : 4.8 miles. What is the actual distance in miles between two points that are 5.5 inches apart on the map?

(1) 1.75
(2) 2.29
(3) 8.3
(4) 13.2
(5) 26.4

5. A person who is 6 feet tall casts a shadow that is 8.5 feet long. At the same time of day, a person who is $4\frac{1}{4}$ feet tall would cast how long of a shadow?

(1) 12.0 feet
(2) 6.75 feet
(3) 6.23 feet
(4) 6.02 feet
(5) 3.0 feet

Questions 6 through 8 refer to the following map.

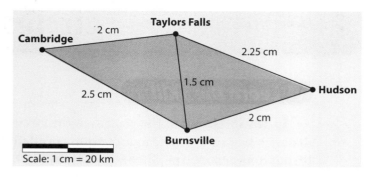

6. What is the actual distance in kilometers between Burnsville and Taylors Falls?

(1) 13.3
(2) 15
(3) 20
(4) 23.3
(5) 30

7. Jack drove from Cambridge to Burnsville. Pedro drove from Hudson to Burnsville. How much farther did Jack drive than Pedro?

(1) 0.5 km
(2) 10 km
(3) 19.5 km
(4) 40 km
(5) Not enough information is given.

8. Carl drives from Cambridge to Taylors Falls and then to Hudson to go to work each day. Each night he drives the same way home to Cambridge. How many kilometers does he commute each workday?

(1) 170
(2) 85
(3) 70
(4) 45
(5) 40

9. A furniture maker made a model of a table design. The model of the table is 12 inches long. The actual table will be 60 inches long. What is the scale of the model?

(1) 6 in. : 1 in.
(2) 1 in. : 4 in.
(3) 1 in. : 5 in.
(4) 1 in. : 12 in.
(5) 6 in. : 36 in.

UNIT 4

Circles

① Learn the Skill

In any **circle**, all of the points on the circle are the same distance to the center of the circle. The distance from the center to any point is called the **radius**. The **diameter** is the distance across a circle through its center. The diameter is always twice the radius. You will use the radius and diameter to solve problems relating to the **circumference** (the distance around) and area of circles.

② Practice the Skill

By mastering the skills involving circles, their parts, and how to use formulas to find circumference and area, you will be able to successfully solve problems on the GED Mathematics Test. Read the examples and strategies below. Then answer the question that follows.

Ⓐ To find the area of a circle, you must know the radius of the circle. In some problems, the diameter will be given instead of the radius. In those cases, divide the diameter by 2 to get the radius. Likewise, multiply the radius by 2 to get the diameter.

Ⓑ To find the circumference of a circle, use the formula $C = \pi d$, where $\pi \approx 3.14$. Circumference is measured in units.

Ⓒ To find the area of a circle, use the formula $A = \pi r^2$. Area is measured in square units.

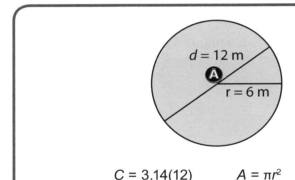

$C = 3.14(12)$
$C = 37.68$ m

$A = \pi r^2$
$A = 3.14(6)^2$
$A = 3.14 \times 6 \times 6$
$A = 113.04$ sq m

☑ TEST-TAKING TIPS

Use estimation to check your answers to problems involving circumference and area. Round 3.14 to 3 before finding the product.

1. The diameter of the smaller circle is equal to the radius of the larger circle, which is 7 inches. What is the area of the larger circle?

   (1) 21.98 in.²
   (2) 49.86 in.²
   (3) 78.86 in.²
   (4) 131.88 in.²
   (5) 153.86 in.²

Directions: Choose the <u>one best answer</u> to each question.

Questions 5 through 7 refer to the following circular area rug.

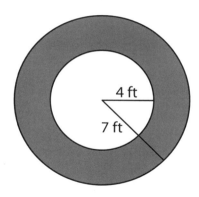

2. Alisha is painting a perfectly round sun as part of a mural on the side of a building. If the diameter of her sun is 15 cm, what is its area in square centimeters?

   (1) 4.78 cm²
   (2) 7.1 cm²
   (3) 176.63 cm²
   (4) 225 cm²
   (5) 706.5 cm²

3. A circle has a diameter of 25 inches. What is its circumference?

   (1) 39.25 inches
   (2) 78.5 inches
   (3) 156.25 inches
   (4) 490.63 inches
   (5) 1,092.5 inches

Questions 4 refers to the information and figure below.

   Jon and Gretchen are laying a circular brick patio in their backyard. The patio is shown in the diagram below.

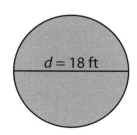

4. If the pavers charge $1.59 per square foot, how much will the pavers charge for the whole patio?

   (1) $56.52
   (2) $89.87
   (3) $404.40
   (4) $1,017.36
   (5) $1,617.60

5. Henry wants to add a fringed edge around the outside of the rug. About how many feet of edging should he buy to go around the outside edge of the rug?

   (1) 69
   (2) 44
   (3) 28
   (4) 25
   (5) 13

6. The rug is divided into a white interior and a gray border. What is the area of the interior of the rug in square feet?

   (1) 5.1
   (2) 16.4
   (3) 28.3
   (4) 50.2
   (5) 153.9

7. What is the area of the entire area rug to the nearest square foot?

   (1) 50
   (2) 103
   (3) 104
   (4) 105
   (5) 154

8. One circle has a radius of 5.5 cm. Another circle has a diameter of 12.5 cm. What is the difference in area between the two circles?

   (1) 4.7 sq cm
   (2) 7 sq cm
   (3) 27.7 sq cm
   (4) 110.7 sq cm
   (5) 395.6 sq cm

UNIT 4

Solid Figures

① *Learn the Skill*

A **solid figure** is a three-dimensional figure. Solid figures include cubes, rectangular solids, pyramids, cylinders, and cones. Problems on the GED Mathematics Test relating to solid figures most often involve volume. You will use your knowledge of algebra and applying formulas to solve these problems.

② *Practice the Skill*

On the GED Mathematics Test, you must be able to recognize solid figures and apply formulas to successfully solve these types of problems. For help with various formulas, refer to p. viii. Read the example and strategies below. Then answer the question that follows.

Ⓐ You previously covered cubes and rectangular solids in Unit 2. Other solid figures you can expect to see on the GED Mathematics Test are square pyramids, cones, and cylinders.

Ⓑ A square pyramid has a square base and four congruent triangular faces. The faces all connect to a single point, called a vertex. The height of a square pyramid forms a right angle with its base.

Ⓒ A cone has a circular base and one vertex. The two are connected by a curved surface. A cylinder has two congruent circular bases connected by a curved surface.

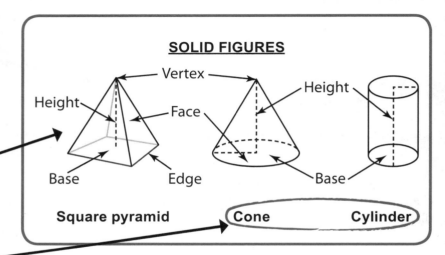

SOLID FIGURES

Vertex · Height · Face · Base · Edge · Base

Square pyramid **Cone** **Cylinder**

☑ TEST-TAKING TIPS

Multiplying by $\frac{1}{3}$ is the same as dividing by 3. For example, to find the volume of a square pyramid, you can find the product of (base edge)^2 × height and then divide by 3.

1. A company sells oatmeal in a cylindrical canister. The canister has a height of 8 inches, and the radius of the base is 3 inches. What is the volume of the container to the nearest cubic inch?

   (1) 24 in.^{3}
   (2) 72 in.^{3}
   (3) 75 in.^{3}
   (4) 226 in.^{3}
   (5) 678 in.^{3}

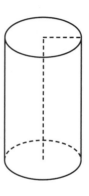

Directions: Choose the <u>one best answer</u> to each question.

Question 2 refers to the following three-dimensional figure.

2. Tracy wants to find out how much ice cream a cone can hold if it is filled exactly to the top.

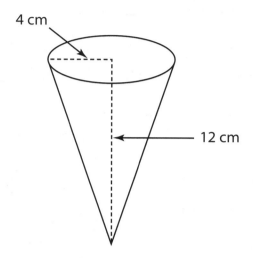

Which expression should Tracy use to find the volume of the cone shown above?

(1) $\frac{1}{3} \times 3.14 \times 4^2 \times 12$

(2) $\frac{1}{3} \times 3.14 \times 8^2 \times 12$

(3) $3.14 \times 4^2 \times 12$

(4) $3 \times 3.14 \times 4^2 \times 12$

(5) $\frac{1}{3} \times 3.14 \times 4 \times 12$

3. A rectangular bale of hay has the following dimensions: length = 40 inches, height = 20 inches, and width = 20 inches. Darla had 50 hay bales delivered to her farm. How many cubic inches of hay did she have delivered?

(1) 4,000 in.³
(2) 8,000 in.³
(3) 16,000 in.³
(4) 160,000 in.³
(5) 800,000 in.³

4. A chocolate shop makes specialty shapes and sizes of chocolate. Lia ordered the two chocolate figures shown below. To find out which contains more chocolate, she finds the volume of each.

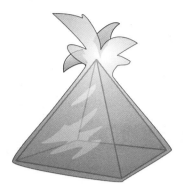

For which two solid figures will she use the formulas for volume?

(1) cylinder and cone
(2) cone and square pyramid
(3) cube and rectangular solid
(4) rectangular solid and cone
(5) square pyramid and cylinder

5. The rain barrel shown below has a volume of 9,156.24 cubic inches.

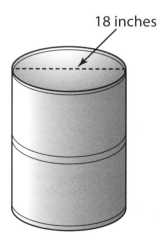

18 inches

What is the height of the rain barrel in inches?

(1) 9
(2) 18
(3) 36
(4) 81
(5) 324

UNIT 4

Irregular Figures

1 Learn the Skill

Most shapes in the real world are not perfect circles or squares. Many figures are **irregular**. They are made up of many plane or solid shapes. To find the area or volume of the combined figure, you must find the area or volume of each part and add them together. You also can find the perimeter of an irregular figure.

2 Practice the Skill

Understanding how to solve problems involving the area, volume, and perimeter of irregular figures will improve your ability to successfully solve real-world area, volume, and perimeter problems on the GED Mathematics Test. Read the example and strategies below. Then answer the question that follows.

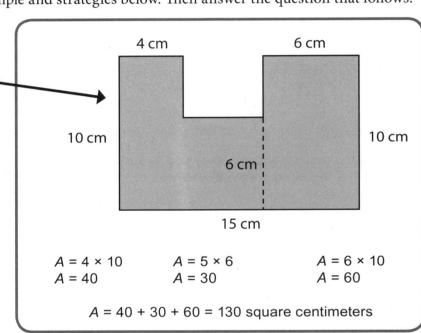

A To find the area of an irregular figure, first divide the figure into simple shapes. This figure can be divided into three rectangles. The dimensions of the two outer rectangles are given. One side of the middle rectangle is given (6 cm). To find the other side, use the measurements of the sides you know. The length of the entire figure is 15 cm. By subtracting 4 cm and then 6 cm, you can find the length of the middle rectangle (5 cm).

TEST-TAKING TIPS

To find the volume of an irregular figure, separate the irregular figure into simple figures. Then find the volume of each simple solid figure, and add to find the total volume.

1. What is the area in square feet of the figure shown below?

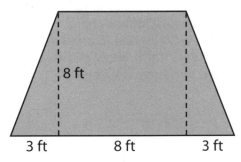

(1) 40
(2) 64
(3) 76
(4) 88
(5) 112

Directions: Choose the <u>one best answer</u> to each question.

2. Kirsten sewed a tablecloth in the shape shown below. What is the area of her tablecloth in square feet?

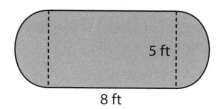

5 ft

8 ft

(1) 40.0
(2) 47.9
(3) 55.7
(4) 56.0
(5) 59.6

<u>Questions 3 and 4</u> refer to the following figure.

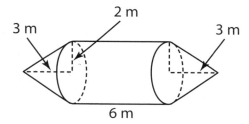

2 m

3 m   3 m

6 m

3. What is the combined volume of the cones in the figure?

(1) 75.36 cubic meters
(2) 37.68 cubic meters
(3) 25.12 cubic meters
(4) 18.84 cubic meters
(5) 12.56 cubic meters

4. What is the volume of the figure to the nearest cubic meter?

(1) 50
(2) 75
(3) 100
(4) 125
(5) 151

<u>Questions 5 through 7</u> refer to the following information and diagram.

Karen and Bill had cement poured to make the patio shown in the diagram.

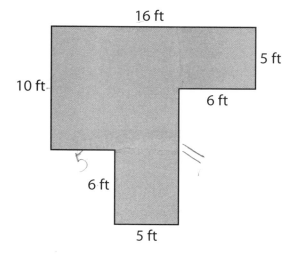

16 ft

5 ft

10 ft

6 ft

6 ft

5 ft

5. What is the area of Karen and Bill's patio in square feet?

(1) 40
(2) 70
(3) 100
(4) 130
(5) 160

6. Karen and Bill are laying a decorative tile border around the outside of the patio. What is the perimeter of the patio in feet?

(1) 8
(2) 64
(3) 96
(4) 192
(5) 768

7. If Karen and Bill have the cement for the patio poured 3 inches deep, how many cubic feet of cement will they use?

(1) 32
(2) 40
(3) 90
(4) 160
(5) 480

UNIT 4

Pythagorean Theorem

① Learn the Skill

A **right triangle** is a triangle with a right (90°) angle. The legs (shorter sides) and **hypotenuse** (longest side) of a right triangle have a special relationship. The **Pythagorean theorem** describes this relationship: In any right triangle, the sum of the squares of the lengths of the legs is equal to the square of the length of the hypotenuse.

② Practice the Skill

Whenever two sides of a right triangle are given, the Pythagorean theorem can be used to find the measure of the third side. Use the Pythagorean theorem for right triangles only. Read the examples and strategies below. Then answer the question that follows.

A The Pythagorean theorem is written:
$$a^2 + b^2 = c^2$$
where a and b are the legs, and c is the hypotenuse. The hypotenuse is always the side opposite the right angle. After solving for the missing side length, check to make sure that the hypotenuse has the greatest measure. Here, solve for c using the values for a and b.

B Use the Pythagorean theorem to find the distance between points A and B. Count units to find the lengths of the legs and use the Pythagorean theorem to solve for the hypotenuse.
$$c^2 = 3^2 + 6^2$$
$$c^2 = 45$$
$$c \approx 6.7$$

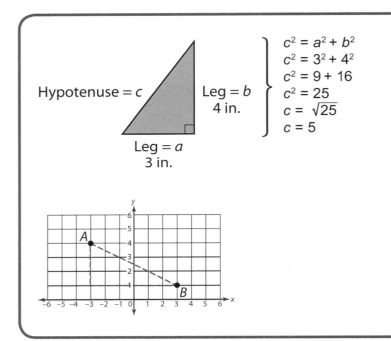

$$c^2 = a^2 + b^2$$
$$c^2 = 3^2 + 4^2$$
$$c^2 = 9 + 16$$
$$c^2 = 25$$
$$c = \sqrt{25}$$
$$c = 5$$

Hypotenuse = c Leg = b
4 in.

Leg = a
3 in.

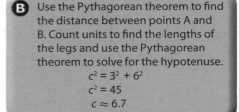

☑ TEST-TAKING TIPS

If the sides of a triangle make the equation $a^2 + b^2 = c^2$ true, then the triangle proves to be a right triangle.

1. The bottom of a ladder is resting 5 feet from the wall of a garage. The wall and the ground form a right angle. If the ladder is 10 feet long, how far up the wall does it reach?

   (1) 5.0 ft
   (2) 6.4 ft
   (3) 7.5 ft
   (4) 8.7 ft
   (5) 11.2 ft

UNIT 4

Directions: Choose the one best answer to each question.

Questions 2 and 3 refer to the following information and diagram.

A telephone pole is 30 feet tall. A cable attached to the top of the pole is anchored to the ground 15 feet away from the base of the pole.

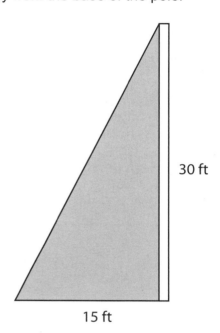

30 ft

15 ft

2. What is the length of the cable to the nearest tenth of a foot?

(1) 26.0
(2) 30.7
(3) 32.2
(4) 33.5
(5) 45.1

3. If a 35-foot cable were run from the top of the pole and anchored to the ground at a distance from the pole, about how far away from the pole would it be anchored?

(1) 16 feet
(2) 18 feet
(3) 30 feet
(4) 38 feet
(5) 46 feet

Questions 4 and 5 refer to the following information and diagram.

The river is 120 meters wide. Sara starts out swimming across the river. The current pushes her, so she ends up 40 meters downriver from where she started.

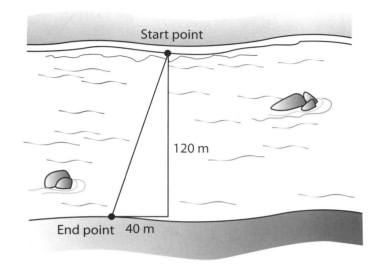

Start point

120 m

End point   40 m

4. To the nearest meter, how many meters did Sara actually swim?

(1) 80
(2) 113
(3) 105
(4) 126
(5) 160

5. If the current had not been so strong and had only swept Sara 20 meters downstream, about how many meters would she actually have swam?

(1) 122
(2) 118
(3) 116
(4) 106
(5) 100

6. What is the distance between points M $(-4, 5)$ and N $(4, 3)$?

(1) 6.3
(2) 6.5
(3) 7.1
(4) 7.3
(5) 8.2

UNIT 4

Unit 4 Review

On the GED Mathematics Test you will be asked to write your answers in different ways. Below are two ways to write your answers for this Unit Review.

Horizontal-response format

①②●④⑤

To record your answers, fill in the numbered circle that corresponds to the answer you select for each question in the Unit Review. Do not rest your pencil on the answer area while considering your answer. Make no stray or unnecessary marks. If you change an answer, erase your first mark completely. Mark only one answer space for each question; multiple answers will be scored as incorrect.

Alternate-response format

To record your answers for an alternate format question
- Begin in any column that will allow your answer to be entered;
- Write your answer in the boxes in the top row;
- In the column beneath a fraction bar or decimal point (if any) and each number in your answer, fill in the bubble representing that character;
- Leave blank any unused column.

<u>Directions</u>: Choose the <u>one best answer</u> to each question.

<u>Questions 1 and 2</u> refer to the following text and figure.

The figure below shows several rays with a common vertex.

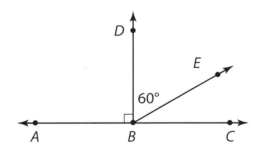

1. Which angles are complementary?

    (1) ∠ABD and ∠DBE
    (2) ∠DBE and ∠DBC
    (3) ∠ABC and ∠EBC
    (4) ∠ABD and ∠EBC
    (5) ∠DBE and ∠EBC

①②③④⑤

2. Which angle has a measure of 150°?

    (1) ∠ABE
    (2) ∠ABD
    (3) ∠DBE
    (4) ∠CBD
    (5) ∠ABC

①②③④⑤

3. The distance between two cities on a map is 8.5 cm. If the map scale is 1 cm : 15 km, what is the actual distance in kilometers between the two cities?

    (1) 127.5
    (2) 63.8
    (3) 31.9
    (4) 3.5
    (5) 1.7

①②③④⑤

Questions 4 and 5 refer to the following figure.

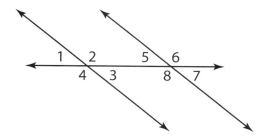

4. What is the value of x?

(1) 30
(2) 45
(3) 60
(4) 90
(5) Not enough information is given.

①②③④⑤

5. What is the value of y?

(1) 60
(2) 90
(3) 100
(4) 120
(5) Not enough information is given.

①②③④⑤

6. A chocolate company sells its specialty hot cocoa in cylindrical canisters. The canister holds 3,740 cubic centimeters of cocoa.

17.5 cm

If the canister is 17.5 cm high, what is the diameter of the canister in centimeters?

(1) 8.25
(2) 16.5
(3) 21.65
(4) 33.0
(5) 68.0

①②③④⑤

Question 7 refers to the figure below.

The figure shows two parallel lines intersected by a transversal.

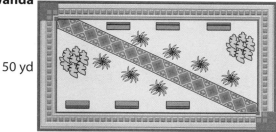

7. Which two angles have the same measure?

(1) ∠4 and ∠2
(2) ∠3 and ∠6
(3) ∠2 and ∠5
(4) ∠7 and ∠4
(5) Not enough information is given.

①②③④⑤

8. Instead of following the sidewalk around the outside of a park to her car, Wanda cut through the park as shown.

Wanda

50 yd

120 yd

Wanda's car

How many fewer yards did Wanda walk than if she had taken the sidewalk back to her car?

(1) 130
(2) 120
(3) 80
(4) 40
(5) 10

①②③④⑤

UNIT 4

Questions 9 and 10 refer to the following diagram.

A human-made backyard pond is shown in the diagram.

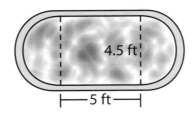

4.5 ft

⊢—5 ft—⊣

9. What is the perimeter of the pond?

(1) 28.26 ft
(2) 24.13 ft
(3) 19.13 ft
(4) 14.13 ft
(5) 13.5 ft

①②③④⑤

10. What is the area of the pond in square feet?

(1) 30.45
(2) 36.63
(3) 38.40
(4) 54.29
(5) 86.09

①②③④⑤

Questions 11 and 12 refer to the following figure.

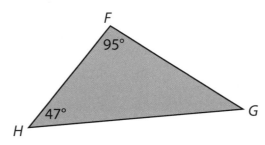

F

95°

47°

H

G

11. What is the measure of ∠G?

(1) 48°
(2) 38°
(3) 28°
(4) 18°
(5) Not enough information is given.

①②③④⑤

12. Which term best describes △HFG?

(1) acute
(2) isosceles
(3) equilateral
(4) obtuse
(5) right

①②③④⑤

Questions 13 and 14 refer to the text and figure below.

Carlos sewed the round picnic blanket shown in the diagram.

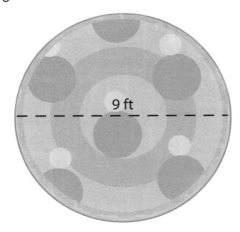

9 ft

13. What is the area of the blanket in square feet?

(1) 21.19
(2) 28.26
(3) 63.59
(4) 127.17
(5) 254.34

①②③④⑤

14. What is the circumference of the blanket in inches?

(1) 63.59
(2) 169.56
(3) 254.34
(4) 339.12
(5) 763.02

①②③④⑤

Unit 4 Review | Geometry

Question 15 refers to the diagram below.

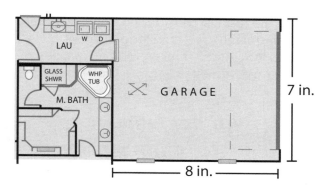

7 in.

8 in.

15. If the scale is 1 inch : 3 feet, what are the actual dimensions of the garage?

(1) 7 feet by 8 feet
(2) 14 feet by 16 feet
(3) 21 feet by 24 feet
(4) 24 feet by 25 feet
(5) 27 feet by 28 feet

①②③④⑤

Question 16 refers to the following text and figure.

Angle C in this parallelogram measures 125°.

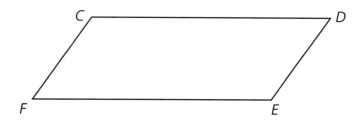

16. What is the measure of ∠D?

(1) 45°
(2) 55°
(3) 65°
(4) 125°
(5) Not enough information is given.

①②③④⑤

Question 17 refers to the following text and figure.

The intersecting lines form a right triangle.

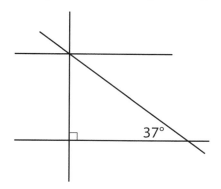

37°

17. What is the measure of the unknown angle of the triangle?

(1) 37°
(2) 45°
(3) 53°
(4) 74°
(5) Not enough information is given.

①②③④⑤

18. A snack shop sells frozen cookie dough in a cone as shown in the diagram.

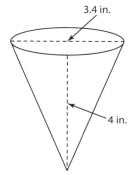

3.4 in.

4 in.

The snack shop buys cookie dough in containers that hold 480 cubic inches of cookie dough. About how many cones can be filled with one container?

(1) about 40
(2) about 36
(3) about 24
(4) about 13
(5) about 10

①②③④⑤

Questions 19 and 20 refer to the following diagram.

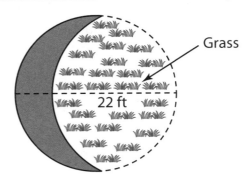

Grass

22 ft

The pool of a fountain in a park is shaped like a crescent moon, as shown in the red part of the diagram.

19. If the grass covers 253.3 square feet, what is the area of the fountain pool?

    (1) 126.64 square feet
    (2) 379.94 square feet
    (3) 759.88 square feet
    (4) 1,013.17 square feet
    (5) 1,266.46 square feet

    ①②③④⑤

20. A circular fence is to be built around the grass and pool area. There needs to be a distance of 5 feet between the fence and the pool and grass. How many feet of fencing will be needed to the nearest tenth?

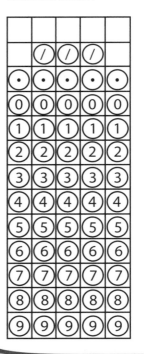

Question 21 refers to the text and figures below.

A corporate farm is considering buying two large plots of farmland. The smaller of the two plots is 5.6 miles long and 3.8 miles wide. The sides of the larger plot are exactly twice the size of the smaller plot.

PLOTS OF FARMLAND

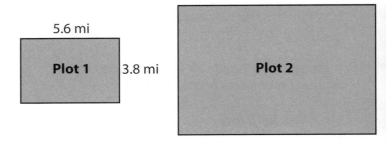

5.6 mi

Plot 1   3.8 mi

Plot 2

21. What is the length of the larger plot of farmland?

Question 22 refers to the following diagram.

The two pyramids in the figure have the same volume.

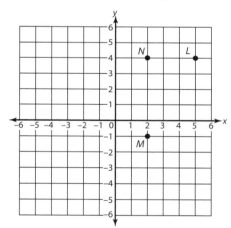

12 cm

12 cm

22 cm

$h = 15$ cm

22. What is the volume of the two square pyramids combined?

(1) 120 cubic centimeters
(2) 720 cubic centimeters
(3) 1,440 cubic centimeters
(4) 2,160 cubic centimeters
(5) 4,320 cubic centimeters

①②③④⑤

Question 23 refers to the following coordinate grid.

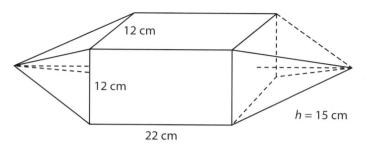

23. What is the distance to the nearest unit between point L and point M?

(1) 3
(2) 4
(3) 5
(4) 6
(5) 7

①②③④⑤

Questions 24 and 25 refer to the following map.

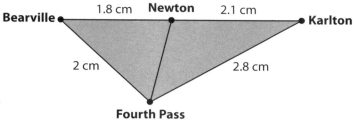

Bearville   1.8 cm   Newton   2.1 cm   Karlton

2 cm   2.8 cm

Fourth Pass

24. If the scale of the map is 2 cm : 29.8 km, what is the actual distance in kilometers between Fourth Pass and Karlton?

(1) 5.6
(2) 10.64
(3) 21.29
(4) 41.72
(5) 83.44

①②③④⑤

25. Lisa drove from Bearville to Newton and then to Karlton. Then she drove from Karlton to Fourth Pass. If the map scale is 1 cm : 35 km, how many kilometers did Lisa drive?

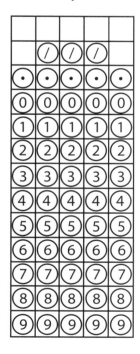

Answer Key

UNIT 1 NUMBER SENSE AND OPERATIONS

LESSON 1, pp. 2–3

1. (5), The 8 in the hundreds place rounds the 6 thousands to 7 thousands.

2. (2), 82 is written as eighty-two.

3. (4), The 6 in the ones place rounds the 8 tens to 9 tens, or 90.

4. (2), 1,384 is greater than 1,337 but less than 1,420.

5. (3), 2,250 has 2 hundreds, 2,450 has 4 hundreds, and 2,700 has 7 hundreds.

6. (1), All numbers have 1 hundred thousand. January and February both have 5 ten thousands, but January has 5 thousands and February has none.

7. (2), January and February have the two highest values, and these months are in the winter.

LESSON 2, pp. 4–5

1. (2), 6 ones minus 0 ones is 6 ones, 5 tens minus 4 tens is 1 ten, and then regroup 1 thousand as 10 hundreds to see that 12 hundreds minus 3 hundreds is 9 hundreds.

2. (5), 7 ones plus 3 ones is 10 ones, or 1 ten, and 6 tens plus 8 tens plus the extra ten is 15 tens, or 1 hundred and 5 tens. 4 tens plus 5 tens plus the extra ten is 10 tens.

3. (2), Regroup 307 as 2 hundreds, 9 tens, and 17 ones, and then subtract 1 ten and 9 ones.

4. (4), 40 hours × $9 per hour = $360 earned in one week.

5. (2), $45 per month × 12 months per year = $540 for one year.

6. (4), $64 divided evenly among 4 people is $16 per person.

7. (2), Claire must find how many groups of 12 are in 504. 504 ÷ 12 = 42 with no remainder.

8. (3), $630 × 18 months = $11,340 in all.

9. (2), 4,000 – 3,518 = 482 yards left, and then 482 yards divided by 2 games = 241 yards.

LESSON 3, pp. 6–7

1. (4), 64 hot dogs × $3 = $192, and 38 yogurts × $2 = $76, so $192 + $76 = $268 in all.

2. (2), A sale price of $42 – a $5 coupon = $37; the original price of $54 is a distraction.

3. (5), $645 + $78 + $25 = $748 in all.

4. (1), 143 students in all – 67 soccer students = 76 non-soccer students; the number of students who signed up to play softball were only part of those playing a sport other than soccer.

5. (3), Multiply 144 by $6 because the store is paying for the shirts; the product is $864.

6. (1), The profit for one men's sport shirt is $16 – $9, which is $7, and 508 shirts × $7 per shirt = $3,556.

7. (2), A budget of $5,500 – $5,000 spent = $500 remaining. Then $500 ÷ 4 = $125.

LESSON 4, pp. 8–9

1. (3), Rename the fractions as $\frac{8}{20}, \frac{15}{20}, \frac{16}{20}$, and $\frac{5}{20}$. The greatest fraction is $\frac{16}{20}$, which corresponds to Ella.

2. (4), Using the common denominator of 30, rename the fractions as $\frac{15}{30}, \frac{30}{30}, \frac{18}{30}, \frac{10}{30}$, and $\frac{24}{30}$. One-half $= \frac{15}{30}$, so the only fraction less than one-half is $\frac{10}{30}$, which corresponds to Team 4.

3. (2), $\frac{1}{1}$ means a full bowl, so Team 2 won the race.

4. (3), The two given denominators are 4 and 8. Since 4 divides evenly into 8, 8 is the lowest common denominator.

5. (5), Clark needs to know how many halves are in $2\frac{1}{2}$. $2\frac{1}{2} = \frac{5}{2}$, so there are 5 halves in $2\frac{1}{2}$.

6. (2), Rename the fractions as $\frac{5}{20}, \frac{8}{20}, \frac{6}{20}$, and $\frac{1}{20}$. Since $\frac{8}{20}$ is the greatest, James won the election.

7. (1), Convert all fractions to mixed numbers. Only 2 friends, Naomi and Kirsten, ate more than 2 slices. Because $\frac{1}{2}$ is larger than $\frac{1}{3}$, Naomi ate the most pizza.

LESSON 5, pp. 10–11

1. (3), To add the fractions, rename them as $\frac{12}{20}$ and $\frac{15}{20}$. Rename the sum of $\frac{27}{20}$ as $1\frac{7}{20}$.

2. (5), To add the fractions, rename them as $3\frac{2}{6}$ and $1\frac{3}{6}$. Add the numerators and whole numbers to find a sum of $4\frac{5}{6}$.

3. (3), To find the difference, solve $5\frac{3}{12} - 2\frac{2}{12}$ to get $3\frac{1}{12}$.

4. (1), To find half of something, multiply by one-half. $1\frac{3}{4} = \frac{7}{4}$, $\frac{7}{4} \times \frac{1}{2} = \frac{7}{8}$.

5. (1), To find the number of curtains, divide the total length by the amount needed for each curtain. $11\frac{1}{4} \div 2\frac{1}{4} = \frac{45}{4} \div \frac{9}{4} = \frac{45}{4} \times \frac{4}{9}$. Simplify by dividing the numerator and denominator by 4 and 9. $\frac{45}{4} \times \frac{4}{9} = \frac{5}{1} = 5$

6. (1), 8 stacks of $\frac{2}{3}$ means $8 \times \frac{2}{3} = \frac{8}{1} \times \frac{2}{3} = \frac{16}{3} = 5\frac{1}{3}$.

7. (2), The sum is $3\frac{3}{8} + 1\frac{2}{8} + 2\frac{4}{8} = 6\frac{9}{8} = 6 + 1\frac{1}{8} = 7\frac{1}{8}$.

8. (3), Rename the fractions using 24 as a denominator. The sum is $2\frac{29}{24}$, which is $2 + 1\frac{5}{24}$, which is $3\frac{5}{24}$.

LESSON 6, pp. 12–13

1. (4), Solve $\frac{3}{\$12} = \frac{5}{x}$ to find that $x = \$20$.

2. (4), To simplify $\frac{92}{64}$, divide the numerator and denominator by 4 to get $\frac{23}{16}$.

3. (3), Solve $\frac{558}{9 \text{ hr}} = \frac{x}{1 \text{ hr}}$ to find that $x = 62$ miles.

4. (1), To simplify $\frac{2}{10}$, divide the numerator and denominator by 2 to get $\frac{1}{5}$.

5. (2), Solve $\frac{4}{20} = \frac{x}{120}$ to find that $x = 24$ miles.

6. (5), Solve $\frac{2}{7} = \frac{14}{x}$ to find that $x = 49$ children.

7. (2), Solve $\frac{65}{1 \text{ hr}} = \frac{260}{x}$ to find that $x = 4$ hours.

8. (5), Solve $\frac{3}{2} = \frac{144}{x}$ to find that $x = 96$ trucks.

9. (2), To simplify $\frac{25}{5}$, divide the numerator and denominator by 5 to get $\frac{5}{1}$.

10. (2), Solve $\frac{1}{12} = \frac{x}{36}$ to find that $x = 3$ teachers.

LESSON 7, *pp. 14–15*

1. (4), To round to the nearest dollar, look at the tenths place. The tenths digit in 30,237.59 is 5, so the 5 rounds 7 ones to 8.

2. (4), To round to the nearest tenth, look at the hundredths digit. Seven hundredths rounds the 2 to a 3.

3. (1), The sum of the prices is $104.46; round down to $104.00 to get to the nearest dollar.

4. (3), All numbers have 3 tenths, but .3304 is the only number with 3 hundredths, so it is the greatest.

5. (1), 1.59 and 1.76 have the least amount of ones. 1.59 has 5 tenths and 1.76 has 7 tenths, so 1.59 is less.

6. (4), There are four numbers less than 2.25: 1.59, 2.07, 1.76, and 2.15.

7. (3), The two greatest numbers are 25.98 and 25.57 because they both have 5 ones. 25.98 has 9 tenths, while 25.57 only has 5 tenths, so 25.98 is greatest. Therefore, 24.30 is the least.

LESSON 8, *pp. 16–17*

1. (1), The cost of the coffee and muffin is $2.95 + $1.29, which is $4.24. Subtract $4.24 from $5.00 to get a difference of $0.76.

2. (4), Multiply $3.50 by 6 to get a product of $21.00.

3. (3), Ben paid $9.32 for bread because $2.33 × 4 = $9.32. He paid $2.58 for butter because $1.29 × 2 = $2.58. The difference of $9.32 − $2.58 = $6.74.

4. (2), The cost of 6 gallons of milk is 6 × $3.50, or $21.00. The cost of 5 boxes of cereal is 5 × $3.85, or $19.25. The sum of $21.00 + $19.25 is $40.25.

5. (3), $33.40 ÷ 4 = $8.35

6. (2), The difference in cost between reams is $5.25 − $3.99 = $1.26, so $1.26 saved per ream × 15 reams = $18.90.

7. (4), Divide the length of the whole rope by the number of equal pieces. 14.4 ÷ 4 = 3.6

8. (5), Coach Steve will spend $17 × 12, or $204, on uniforms. He will spend $12.95 × 6, or $77.70, on soccer balls. The sum of the two products is $281.70.

9. (3), Coach Steve will spend $10.95 × 12, or $131.40, on shin guards. He will spend $8.95 × 12, or $107.40, on kneepads. The difference of $131.40 − $107.40 is $24.00.

LESSON 9, *pp. 18–19*

1. (4), $\frac{27}{45} = 27 \div 45 = 0.6$, and $0.6 = 60\%$.

2. (4), $25\% = \frac{25}{100} = \frac{25 \div 25}{100 \div 25} = \frac{1}{4}$.

3. (2), $7\frac{1}{2} = 7\frac{5}{10} = 7.5$.

4. (1), $\frac{3}{8} = 3 \div 8 = 0.375$, and $0.375 = 37.5\%$.

5. (3), $\frac{1}{8} = 1 \div 8 = 0.125$.

6. (4), Since $0.22 = \frac{22}{100}$, 22 people said yes. Since 22 people said yes, 100 − 22, or 78, people said no. $\frac{78}{100} = \frac{78 \div 2}{100 \div 2} = \frac{39}{50}$.

7. (3), $\frac{9}{13} = 9 \div 13 \approx 0.6923$, and $0.6923 = 69.23\%$, which rounds to 69.2%.

8. (5), Either find the product of 0.88 × 25 or solve $\frac{88}{100} = \frac{x}{25}$ to get 22 questions.

9. (4), 75% of 300 is the same as $\frac{3}{4} \times 300$, which equals 225.

LESSON 10, *pp. 20–21*

1. (2), Using the formula $I = prt$, $I = 1{,}000(0.05)\left(\frac{1}{2}\right) = 25$, which means Kirsten owes $25 in interest. She owes $25 plus the $1,000 she borrowed, which equals $1,025.

2. (4), Solve $\frac{12}{100} = \frac{x}{552}$ to find that $x = \$66.24$.

3. (4), Divide 309 by 824 to get 0.375. Then multiply by 100 to get 37.5%.

4. (3), Andrew received a raise of $25,317.40 − $24,580.00, or $737.40. Then solve $\frac{737.40}{24{,}580} = \frac{x}{100}$ to find that $x = 3\%$.

5. (4), Multiply $425 by 0.06 to find the tax, which is $25.50. Then add $425 + $25.50 to find a total of $450.50.

6. (2), Multiply 0.45 by 420 to get 189.

7. (2), To find the discount, multiply 0.20 by $659 to get $131.80. Then subtract $659.00 − $131.80 to get a sale price of $527.20.

8. (5), Using the formula $I = prt$, $I = 5{,}000(0.05)\left(\frac{9}{12}\right) = 187.50$.

9. (2), Multiply 0.20 by $2,250 to get a product of $450.

10. (1), Multiply 0.40 by 25 to find that 10 new employees will be hired. 25 current employees + 10 new employees = 35 employees in all.

UNIT 1 REVIEW, *pp. 22–27*

1. (5), If 35% were in favor, then 100% − 35% = 65% objected. 0.65 × 1,200 = 780.

2. (4), The chairs cost $65.30 × 4, or $261.20. The total cost is $261.20 + $764.50 = $1,025.70.

3. (2), Subtract 210.5 − 135.8 to get a difference of 74.7.

4. (3), Divide $4\frac{1}{2}$ by $1\frac{1}{2}$: $4\frac{1}{2} \div 1\frac{1}{2} = \frac{9}{2} \div \frac{3}{2} = \frac{9}{2} \times \frac{2}{3} = 3$.

5. (4), To find the total number of students, add 118 + 54 + 468 + 224 to get a sum of 864. Simplify $\frac{54}{864} : \frac{54 \div 54}{864 \div 54} = \frac{1}{16}$.

6. (4), Add 468 + 224 to get a sum of 692. Then simplify $\frac{692}{864}$ by dividing the numerator and denominator by 4 to get $\frac{173}{216}$.

7. (5), Using the formula $I = prt$, $I = 1{,}250(0.06)(3) = 225$. Kara received $225 in interest plus the original $1,250 she invested, which equals $1,475.

8. (7/12), Use a common denominator of 12. To solve $5\frac{4}{12} - 4\frac{9}{12}$, rename the mixed numbers as improper fractions: $\frac{64}{12} - \frac{57}{12} = \frac{7}{12}$.

9. (1), The pretzels cost $1.95 × 2, or $3.90. The soft drinks cost $0.99 × 2, or $1.98. The pretzels and soft drinks cost $3.90 + $1.98, or $5.88 in all. Subtract $10.00 − $5.88 to find a difference of $4.12.

10. (3), Rename the fractions as $\frac{10}{60}, \frac{3}{60}, \frac{20}{60}, \frac{12}{60}$, and $\frac{15}{60}$. Since $\frac{20}{60}$ is the greatest, soccer has the greatest participation.

11. (3), Add $\frac{1}{4} + \frac{1}{6} = \frac{3}{12} + \frac{2}{12} = \frac{5}{12}$.

ANSWER KEY

UNIT 1 (continued)

12. (7/102), Simplify $\frac{28}{408}$ by dividing the numerator and denominator by 4.

13. (5), Solve $\frac{301.5}{4.5 \text{ hr}} = \frac{x}{1 \text{ hr}}$ to find that $x = 67$ miles.

14. (2), The profit for one share is $52 – $43, or $9. Multiply $9 by 20 shares to get a product of $180.

15. (4), 426 people divided by 65 per bus is 6 buses, with 36 people that cannot fit in a bus. One more bus is needed, so $6 + 1 = 7$ buses are needed in all.

16. (3), Fifty-six thousand is written as 56,000, and two hundred twenty-eight is written as 228.

17. (298), Delaney has $198 + $226, which is $444. She wrote checks for $54 and $92, which is $146, so $444 – $146 = $298.

18. (2), 9.65 and 9.19 both have 9 ones; 9.65 has 6 tenths, but 9.19 only has 1 tenth, so 9.65 is the greatest.

19. (2), The kid's platters cost $3 \times \$3.50$, or $10.50. Add $10.50 + $9.19 + $5.89 to get a sum of $25.58. Subtract $50.00 – $25.58 to find a difference of $24.42.

20. (4), The elk sandwiches cost $\$9.65 \times 2$, or $19.30. The kid's platters cost $\$3.50 \times 3$, or $10.50. The difference of $19.30 – $10.50 = $8.80.

21. (3), Solve $\frac{2}{3} = \frac{x}{180}$ to find that $x = 120$.

22. (1), Find the number of groups of $1\frac{2}{3}$ in 4 hours. $\frac{4}{1} \div 1\frac{2}{3} = \frac{4}{1} \div \frac{5}{3} = \frac{4}{1} \times \frac{3}{5} = \frac{12}{5}$, and $\frac{12}{5} = 2\frac{2}{5}$.

23. (4), Multiply 0.84 by 175 to get a product of 147.

24. (5), Multiply: $\frac{2}{3} \times \frac{24}{1} = \frac{2}{1} \times \frac{8}{1} = 16$.

25. (1), Simplify $\frac{10}{2}$ by dividing the numerator and denominator by 2 to get $\frac{5}{1}$.

26. (2), Solve $\frac{78}{200} = \frac{x}{400}$ to find that $x = 156$.

27. (6), The population changed by $45,687 – 43,209$, or 2,478, people. Solve $\frac{2,478}{43,209} = \frac{x}{100}$ to find that $x \approx 5.73$, which rounds to 6.

28. (1), $54\% = \frac{54}{100}$, which simplifies to $\frac{27}{50}$.

29. (4), Twelve months $\times$ $165.40 = $1,984.80.

30. (3), Multiply by 3: $1\frac{3}{8} \times \frac{3}{1} = \frac{11}{8} \times \frac{3}{1} = \frac{33}{8}$, which simplifies to $4\frac{1}{8}$.

31. (5), Solve $\frac{\$8.99}{1} = \frac{x}{1.76}$ to find that $x \approx 15.822$, which rounds to 15.82.

32. (2), $25\frac{4}{5} = 25\frac{8}{10}$, which equals 25.8. Subtract $32.95 – 25.8$ to find a difference of 7.15.

UNIT 2 MEASUREMENT/DATA ANALYSIS

LESSON 1, pp. 30–31

1. (1), Convert 30 mL to 3 cL. There are $3 + 2 = 5$ cL altogether.

2. (4), Convert 6 yd to 18 ft. Samantha needs $18 + 12 = 30$ ft of wood.

3. (5), Convert 2 km to 2,000 m. Jason ran $2,000 + (2)(1,500) + (5)(100) = 2,000 + 3,000 + 500 = 5,500$ m over the two days.

4. (5), Convert 1 cup to 8 fl oz and 1 pint to 16 fl oz. Mr. Trask needs $(2)(6) + 8 + 16 = 12 + 8 + 16 = 36$ fl oz of food.

5. (1), Divide the amounts in milligrams by 10. The students use $2 + 0.5 + (2)(1.5) + 0.5 = 6$ cg of Chemical A altogether.

6. (5), Shantell had 60 cg = 600 mg of Chemical C. She had $600 – 5 = 595$ mg more of Chemical C.

7. (3), Convert 50 cg to 500 mg and 1 g to 1,000 mg. Diego used $15 + 500 + 1,000 = 1,515$ mg of chemicals. He used $1,515 – 535 = 980$ mg more than Dana.

LESSON 2, pp. 32–33

1. (3), The diameter of the fence will be $(25)(2) = 50$ m. The circumference will be $(3.14)(50) = 157$ m.

2. (3), The perimeter of the square is $(12)(4) = 48$ in.

3. (5), The perimeter of the long rectangle is $18 + 2 + 18 + 2 = 40$ in.

4. (5), Some lengths are unknown, so there is not enough information.

5. (4), The perimeter of the short rectangle is $3 + 8 + 3 + 8 = 22$ in.

6. (5), The perimeter of Parcel A is $340 + 250 + 340 + 250 = 1,180$ ft.

7. (3), Divide 340 by 2 to find 170 ft. The perimeter of Parcel C is $170 + 250 + 170 + 250 = 840$ ft.

8. (1), The addition of Parcel D would increase the overall perimeter of the land.

9. (2), Solve $C = 3.14d$ for d, where $C = 56.5$. The diameter is $d = 17.99 \approx 18$ cm.

LESSON 3, pp. 34–35

1. (4), The area of the front wall with the door is $(21)(44) = 924$ ft. The area of the metal door is $(12)(12) = 144$ ft. The area that will need to be painted is $924 – 144 = 780$ ft.

2. (4), Melanie will need $(12)(18) = 216$ ft² of flooring.

3. (5), The area of the bedroom floor is $(12)(18) = 216$ ft², and the area of the closet is $(7)(6) = 42$ ft². It will cost $(\$6)(216 + 42) = \$1,548$ to cover both.

4. (4), It will cost $(\$6)(12)(18) = \$1,296$ to install flooring in the bedroom only.

5. (4), The area of a red strip is $(7)(18) = 126$ sq cm. The area of all of the red strips is $(3)(126) = 378$ sq cm.

6. (1), The area of the dog pen is $(.5)(12)(12.5) = 75$ sq m.

LESSON 4, pp. 36–37

1. (5), The volume of the fish tank is $(40)(15)(25) = 15,000$ cm³.

2. (2), To find the space inside the containers, Manny should calculate the volume.

3. (4), The volume of Container 1 is $(8)(6)(10) = 480$ in³.

4. (5), The volume of Container 2 is $(8)(8)(8) = 512$ in³.

5. (5), The width is needed to find the volume. There is not enough information.

6. (4), The volume of the loft will decrease $(5)(15)(12) = 900$ ft³ after the closet is built.

LESSON 5, pp. 38–39

1. (3), The median height of the runners is $\frac{65 + 67}{2} = 66$ in.

2. (2), The range of the runners' times is $17.2 – 11.8 = 5.4$ seconds.

3. (3), The median time in the race is $\frac{12.8 + 13.5}{2} = 13.15$ seconds.

4. (2), The mean time of the runners is

$$\frac{13.5 + 16 + 12.6 + 15.2 + 12.8 + 11.8 + 17.2 + 12.1}{8}$$

= 13.9 seconds. The difference between Sarah's time and the mean time is $13.9 – 12.1 = 1.8$ seconds.

UNIT 2 (continued)

5. (3), The median is $\frac{22 + 24}{2} = 23$ milk shakes sold. The median is best to use when there is an extreme value (85).

6. (2), Solve $\frac{\$5,229 + \$3,598 + \$6,055 + \$3,110 + \$3,765 + x}{6} = 3,743.14$ for x to find that $x = \$701.84$.

LESSON 6, pp. 40–41

1. (2), By the fifth event, there are 4 striped + 2 black = 6 marbles left in the bag. The probability of selecting a black marble is 2:6 = 1:3.

2. (1), The probability of spinning a 6 out of 8 choices is 1:8.

3. (3), The probability of spinning a 4 or 8 out of 8 choices is $\frac{2}{8} = \frac{1}{4} = .25$.

4. (5), Neither 4 nor 6 are odd, so her probability of spinning an odd number is $\frac{0}{2}$.

5. (1), The probability of receiving a clothing department complaint is $\frac{3}{6 + 4 + 2 + 3} = \frac{3}{15} = \frac{1}{5} = 20\%$.

6. (4), The probability that the next complaint will concern electronics or housewares is $\frac{6 + 4}{6 + 4 + 2 + 3} = \frac{10}{15} = \frac{2}{3}$.

7. (3), The probability it will not rain tomorrow is $\frac{60}{100} = \frac{3}{5}$.

LESSON 7, pp. 42–43

1. (3), The age group 36–45 has the most viewers between the hours of 7 P.M. and 11 P.M.

2. (5), Mrs. Cappelli drove 274 + 203 = 477 miles on Tuesday.

3. (2), Mrs. Cappelli drove 169 + 83 = 252 miles on Wednesday.

4. (4), The greatest increase was $0.36, which occurred between Weeks 4 and 5.

5. (3), The gas price of $2.89 is the lowest over the eight-week period.

LESSON 8, pp. 44–45

1. (2), The difference in rainfall between the two parks was greatest in April.

2. (1), The winner jumped 20 ft, and Contestant A jumped 10 ft.

3. (3), Bars B and E are the only bars with the same length, so Katie and Alana both jumped 15 ft.

4. (5), The range of scores is 20 − 5 = 15.

5. (5), When $y = 80$, $x =$ about 8.

6. (2), In general, as x increases, y also increases.

LESSON 9, pp. 46–47

1. (4), The food section shows about 30%.

2. (5), Slightly more than 50% of the employees drive to work, and the only answer choice greater than 50% is 60%.

3. (2), A significant increase in the price of gasoline likely would result in more employees walking or taking mass transit to work.

4. (5), Poetry has a smaller section than crime.

5. (1), Since they are the most popular categories of books in September, a librarian could make the best argument to order nonfiction and mystery titles.

6. (4), The romance section represents about 15%. So, 30,000 × 0.15 = 4,500 books.

LESSON 10, pp. 48–49

1. (3), Solve $d = rt$ for t using $d = 155$ and $r = 65$. Cheryl should reach her mother's house in $t = 2$ hours and 23 minutes. So, 1:00 + 2 hrs 23 mins = 3:23 ≈ 3:25.

2. (4), The next train leaves at 2:35 P.M., which is 2 hours and 43 minutes later.

3. (3), Add 1 hour and 45 minutes to 3:20 to find that he will arrive at 5:05 P.M.

4. (4), Mindy's time was 34.9 − 34.4 = 0.5 minute, or 30 seconds, faster than Sarah's time.

5. (2), The time passed between 11:35 P.M. and 7:05 A.M. is 7 hours and 30 minutes.

6. (4), Solve $d = rt$ for r with $d = 462$ and $t = 7.5$ to find that $r = 61.6$ mph.

7. (5), Subtract 3 hrs 16 min 6 sec from 3 hrs 20 min 10 sec to find that the winning time was 4 min 4 sec faster than the second-fastest time.

UNIT 2 REVIEW, pp. 50–55

1. (2), Divide 840 ft by 3 to find 280 yards.

2. (4), The mass of two textbooks is 2 kg, which is equal to 2,000 g. So, 2,000 shoelaces would be needed.

3. (4), The perimeter of the two plots is 2(28 + 28 + 15.5) = 2(71.5) = 143 m.

4. Solve $C = 3.14d$ for C with $d = 30$ to find that the circumference is 94.2 inches.

5. (4), The area of the parallelogram is (10.5)(15) = 157.5 ft².

6. (3), The area of one of the triangles is $(\frac{1}{2})(12)(15) = 90$ ft².

7. (3), The area of the dividing wall is (15)(12 + 10.5) = 337.5 ft².

8. The mean number of hours is $\frac{21.5 + 28 + 15.5 + 23 + 29 + 34 + 27 + 35}{8} = 26.625 \approx 26.63$.

9. (1), The median is $\frac{27 + 28}{2} = 27.5$, and the mean is 26.63, so the median is slightly greater than the mean.

10. (5), Container A can hold (8)(8)(8) = 512 cm³.

11. (1), Henry will use (8)(8)(8) + (12)(8)(10) = 1,472 cm³.

12. There are two ways of rolling a 2 or 4, and $\frac{2}{6} = \frac{1}{3}$.

13. (3), The probability of rolling an even number is $\frac{3}{6} = 50\%$.

14. (1), Jane will need $\frac{250 + 250 + 300 + 375}{1000} = 1.175$ kg of powder altogether.

15. (5), The contractors will need 150 + 140 + 150 + 140 = 580 ft of fencing.

16. (2), The greatest jump is shown between 2004 and 2005.

17. (5), The amount of bonuses awarded in 2008 was between $4,000 and $5,000.

18. Solve $d = rt$ for d when $r = \frac{45}{60}$ and $t = 45$ minutes. Devaughn can travel 33.75 miles in 45 minutes.

19. (2), Solve $d = rt$ for t when $d = 850$ and $r = 500$ to find that $t = 1.7$ hours = 1 hour 42 minutes. The plane will arrive at Chicago at (11:30 + 1:42) − 1 hour (time change) = 12:12 P.M.

UNIT 2 *(continued)*

20. (5), The rate is unknown, so there is not enough information.

21. (2), Kim should calculate the area because area covers a flat space.

22. (1), The new volume would be $(2)(15) \times (2)(3) \times (2)(8) = (30)(6)(16) = 2{,}880$ ft³.

23. The chances of a player not winning are $\dfrac{4 \text{ "Sorry"}}{8 \text{ in all}} = \dfrac{1}{2} = 0.5$.

24. (3), The time passed between 11:50 A.M. and 2:10 P.M. is 2 hours and 20 minutes.

UNIT 3 ALGEBRA, FUNCTIONS, AND PATTERNS

LESSON 1, *pp. 58–59*

1. (5), The change in temperature was $12 - (-3) = 12 + 3 = 15°F$.

2. (2), Her net gain is $3 - 4 + 8 = -1 + 8 = 7$ spaces forward.

3. (4), Uyen's new balance is $154 - 40 = 114$.

4. (1), Since Sasha descended, her change in position is $-212 + (-80) = -292$ ft.

5. (5), Tyler's meeting is on floor $6 - 2 + 4 = 8$.

6. (4), There were $3{,}342 - 587 - 32 + 645 = 3{,}368$ students enrolled in the fall.

7. (3), The change in the number of students between May and the fall is $-587 - 32 + 645 = -619 + 645 = 26$.

8. (4), Melanie's score was $8 - 6 - 4 + 3 + 4 = 5$.

LESSON 2, *pp. 60–61*

1. (1), If Gabe's sister's age $= x$, then Gabe's age $= 3x$.

2. (4), Kevin's yard is twice g increased by 10, which is the same as $2g + 10$.

3. (3), The number of employees that work in manufacturing is 500 less than $3s$, which is the same as $3s - 500$.

4. (1), Let Michael's science score $= s$; then his math score $= 8 + \frac{1}{2}s = 8 + \frac{s}{2}$.

5. (5), Julie left $\frac{1}{6}(48) + 2 = 8 + 2 = \$10$ for a tip.

6. (4), An adult ticket costs $2(\$12) - \$4 = \$24 - \$4 = \$20$.

7. (5), The perimeter of the rectangle is represented by $2(2w - 3) + 2(w) = 4w - 6 + 2w = 6w - 6$.

8. (2), Since $A = lw$, the area of the rectangle is represented by $(2w - 3)\,w$.

LESSON 3, *pp. 62–63*

1. (5), The first bill $= x$ and the second bill $= 2x + 5$. Solve $x + 2x + 5 = 157$ for x to find the amount of the first bill.

2. (2), Let $x =$ the first integer and $x + 1 =$ the second integer. Then, $x + x + 1 = 15$ can be used to find the first integer.

3. (1), Let $x =$ the number of literature classes and $2x =$ the number of science classes. Then, $x + 2x = 3$, which is the same as $3x = 3$.

4. (3), Stephanie's age $= 3 + 0.5x$. Since $x = 24$, Stephanie is 15 years old.

5. (5), Let $v = 24$; solve $c = 2 + \frac{1}{3}v$ for c to find that there are 10 cellos in the orchestra.

6. (1), Let $x =$ number of $5 bills and $20 - x =$ number of $1 bills. Solve $5x + 20 - x = 52$ for x to find that there are eight $5 bills.

7. (2), Solve $4x = 2x - 4$ for x to find that $x = -2$.

8. (2), Republican pins $= r$ and Democratic pins $= 3r - 14$. Solve $r + 3r - 14 = 98$ for r to find that there are 28 Republican pins.

9. (1), Add all three sides, $2a - 1 + a + 2a = 5a - 1$, and set equal to 16.5. Solve the equation $5a - 1 = 16.5$ for a.

LESSON 4, *pp. 64–65*

1. (2), Move the decimal point seven places to the left to find 5.8×10^7.

2. (5), Carlos completed $4 \times 4 \times 4 = 64$ squats.

3. (2), Move the decimal point seven places to the left to find 2.54×10^7.

4. (4), The length of the square is $\sqrt{81} = 9.0$ m.

5. (1), The solution is $\dfrac{\sqrt{64}}{4} = \dfrac{8}{4} = 2$.

6. (2), The area of the rectangle is $2^6 \cdot 2^5 = 2^{11}$.

7. (4), The width is $(1.5 \times 10^{-3})(2.0 \times 10^5) = (1.5 \times 2.0)(10^{-3} \times 10^5) = 3.0 \times 10^2$ cm.

8. (3), The expression $5^1 + 4^0$ is equal to $5 + 1 = 6$.

9. (3), Solve $x^2 = 30$ by taking the square root of both sides, so $\sqrt{30} \approx 5.477 \approx 5.5$.

10. (3), The length of one side is $\sqrt{50} \approx 7$ft, so the perimeter is $7 \times 4 = 28$ ft.

LESSON 5, *pp. 66–67*

1. (3), Solve $4^2 - 5 = 16 - 5 = 11$.

2. (3), The pattern 2, 4, 16, 256, 65,536,… is the same as $2, 2^2, 4^2, 16^2, 256^2,…$. so, the rule is to square the previous term.

3. (1), The next term in the pattern is $65{,}536^2 = 4{,}294{,}967{,}296$.

4. (4), A multiple of 5 will result in a whole number output, so $x = 25$.

5. (2), Solve $1{,}035 = 230t$ for t to find 4.5 hours.

6. (4), When x increases by 5, y increases by $0.40, so $1.60 + $0.40 = 2.00.

7. (3), Each term is divided by 2. The fifth term is $24 \div 2 = 12$, and the sixth term is $12 \div 2 = 6$.

8. (3), To find y, multiply x by 5 and then subtract 1, so $4(5) - 1 = 19$.

9. (1), If $f(x) = 4$, then $4 = 2 - \frac{2}{3}x$. Solve for x to find that $x = -3$.

LESSON 6, *pp. 68–69*

1. (1), Factor $x^2 + 5x - 6$ to find $(x + 6)(x - 1)$.

2. (4), The product of $(x + 5)(x - 7)$ is $x^2 + 5x - 7x - 35 = x^2 - 2x - 35$.

3. (1), The product of $(x - 3)(x - 3)$ is $x^2 - 3x - 3x + 9 = x^2 - 6x + 9$.

4. (4), Factor $x^2 - 6x - 16$ to find $(x + 2)(x - 8)$.

5. (1), The area of the rectangle is $(2x - 5)(-4x + 1) = -8x^2 + 22x - 5$.

6. (2), The term $(x + 3)$ is multiplied by $(4x + 1)$ to find $4x^2 + 13x + 3$.

7. (5), Divide both sides by 2, and then factor to find $(x + 6)(x + 3)$. Solve for x to find that $x = -6$ and $x = -3$.

8. (2), Solve $w^2 - 12w + 32$ for x to find that one of the widths is 4 m.

9. (3), The product of $(x - 7)(x + 7)$ is $x^2 - 49$.

UNIT 3 *(continued)*

LESSON 7, *pp. 70–71*

1. (5), Solve the inequality $5x \le 2x + 9$ for x to find $x \le 3$.

2. (4), Solve $x + 5 > 4$ for x to find $x > -1$.

3. (2), The number line shows that x is equal to or less than -2, so $x \le -2$.

4. (4), Let $5x$ = the product of a number and 5. The inequality is $5x + 3 \le 13$.

5. (5), The relationship can be represented by $80 \ge w(3w - 3)$.

6. (1), Kara and Brett have a total of $15 + $22 = $37 for tickets, so $37 < x$.

7. (4), The taxicab charges $2 + $0.50x$, where x = number of miles. Solve $2 + 0.50x \le 8$ for x to find that Josie can ride 12 miles.

8. (2), The relationship can be represented by $x + 12 \le 5x + 3$.

9. (2), Solve $8 - 3x > 2x - 2$ for x to find that $x < 2$.

10. (4), Solve $-5x > 30 - 3(x + 8)$ for x to find that $x < -3$.

LESSON 8, *pp. 72–73*

1. (1), Point C is 2 units right and 2 units up, so the coordinates are (2, 2).

2. (2), Point T is 4 units right and 4 units down, so the coordinates are (4, −4).

3. (2), Point S is 1 unit right and 0 units up or down, so the coordinates are (1, 0).

4. (1), Point P is 5 units left and 5 units down, so the coordinates are (−5, −5).

5. (5), Point D is 3 units right and 5 units down, so the coordinates are (3, −5).

6. (3), Point C is (−2, 3). The new C is 2 units right and 3 units up, so the new coordinates are (2, 3).

7. (1), Point B is (−2, 5). The y-coordinate would decrease by 3, so (−2, 2).

LESSON 9, *pp. 74–75*

1. (2), Let $x = 3$ and $y = -1$; then solve $2(3) + (-1) = 5$ to find that $5 = 5$, so the solution is (3, −1).

2. (3), Let $x = 0$ and $y = 2$; then solve $0 + 2(2) = 4$ to find that $4 = 4$, so the solution is (0, 2).

3. (1), Let $x = 0$ and $y = 0$; then solve $2(0) - (0) = 0$ to find that $0 = 0$, so the solution is (0, 0).

4. (4), Solve $3 = 2x + 2$ for x to find that $x = \frac{1}{2}$.

5. (3), Use the distance formula. Solve $\sqrt{(-4 - 0)^2 + (3 - 0)^2}$ to find $\sqrt{16 + 9} = \sqrt{25} = 5$.

6. (4), Let $x = -5$ and $y = 1$; then solve $-5 + 2(1) = -3$ to find that $-3 = -3$, so the solution is (−5, 1).

7. (5), Use the distance formula. Solve $\sqrt{(2 - 4)^2 + (5 - 3)^2}$ to find $\sqrt{4 + 4} = \sqrt{8} \approx 2.83$.

8. (3), The distance from (−5, 2) to (−3, 1) is $\sqrt{(-5 + 3)^2 + (2 - 1)^2} \approx 2.236$. The distance from (−3, 1) to (−1, −4) is $\sqrt{(-3 + 1)^2 + (1 + 4)^2} \approx 5.385$. Marvin's total distance was $2.236 + 5.385 = 7.621 \approx 7.62$.

LESSON 10, *pp. 76–77*

1. (3), Use the slope formula to find $\frac{4 - 3}{1 - (-1)} = \frac{1}{2}$.

2. (1), Pick two points and use the slope formula to find $\frac{5 - 2}{1 - 0} = \frac{3}{1} = 3$.

3. (4), Use $y = mx + b$ with $m = 3$. Use any point's coordinates to solve for b, such as $2 = 3(0) + b$, so $b = 2$. The formula is $y = 3x + 2$.

4. (3), The slope is $\frac{\text{rise}}{\text{run}} = \frac{2}{32} = \frac{1}{16}$.

5. (3), If f$(x) = 2$, then, using form $y = mx + b$, f$(x) = 2 + 0x$, so the slope is 0.

6. (5), When the equation is written in slope-intercept form, $y = -2x + 4$, we can see the slope is −2. Since choice 5 has a slope of −2, they are parallel.

UNIT 3 REVIEW, *pp. 78–83*

1. (1), The painter earns $20 per hour, and the assistant earns $15 per hour. If h = number of hours, then $20h + 15(h + 5) = 355$.

2. (4), Since $x^2 = 36$, $\sqrt{x} = 6$ or −6. So, $2(6 + 5) = 22$.

3. (3), Sara's new balance is $1,244 + $287 − $50 = $1,481.

4. (2), Subtract 0.5 from the previous term 0 to find the next term is −0.5.

5. (2), Let w = the number of women. The number of men is represented by $5 + \frac{1}{2}w$ or $\frac{1}{2}w + 5$.

6. (0.38), Solve $3x + 0.15 = 1.29$ for x to find $x = 0.38$.

7. (5), The slope of the line is $\frac{2 - 0}{3 - 0} = \frac{2}{3}$.

8. (3), Use point (0, 0) to find $b = 0$. The equation of line Z is $y = \frac{2}{3}x$.

9. (2), Let x = the number of people under 25. The number of people over 25 is represented by $2x - 56$.

10. (4), Solve $5 + 1.25x \le 65$ for x to find that $x \le 48$.

11. (3), Move the decimal point eight places to the left to find 1.496×10^8.

12. The point will be 3 units to the right of the y-axis, so the new coordinates are (3, −2).

13. (3), The slope of $\overline{JL}$ is $\frac{-4 + 4}{-1 + 5} = \frac{0}{4} = 0$.

14. (4), Solve $x(x + 1) = 19 + (x + x + 1)$ for x to find that $x = -4$ or $x = 5$. The integers are negative, so the two integers are −4, and $-4 + 1 = -3$. The greater integer is −3.

15. The distance between the two points is $\sqrt{(-5 - 0)^2 + (4 - 1)^2} = \sqrt{34} = 5.83$.

16. (1), The change in Emmit's account is $-$64 \times 3 = -192.

17. (4), Solve $y = \frac{3}{4}(2)$ to find that the missing number is $\frac{3}{2}$.

18. (4), His new position is $786 - 137 + 542 = +1,191$ feet.

19. (1), The number line shows that x is greater than or equal to 1, so $x \ge 1$.

20. ($255), The cost for one child is $\frac{1}{2}(230) - 30 = $85. The cost for three children is $85 \times 3 = $255.

21. Since side JK is 4 units long, the opposite side must be 4 units long as well. So, the coordinates for the fourth point will be (2, 1).

22. (3), Use points (−2, 1) and (2, 5) to find $m = 1$ and $b = 3$. The equation of this line would be $y = x + 3$.

23. (2), Solve $0 = -16t^2 - 48t + 160$ for t to find that $t = -5$ or $t = 2$. So, the ball takes 2 seconds to reach the ground.

24. (1), Let c = the calf's weight; then the mother's weight is $4c + 200$.

UNIT 3 (continued)

25. (3), Keenan paid $8x + 4.16 = \$73.36$ for the lights. Solve for x to find that each light cost $8.65.

26. (2), There are $8 \cdot 8 \cdot 8 \cdot 8 = 4{,}096$ students at the university.

27. (4), Let x = the number of T-shirts; then the number of T-shirts he can buy is represented by $12 + 32x \leq 80$.

28. (2), Line R is horizontal, so $m = 0$, and the equation is $y = 2$.

29. (1), Using $(0, -2)$ and $(-2, 1)$, $m = \dfrac{1 - (-2)}{-2 - 0} = \dfrac{3}{-2} = -\dfrac{3}{2}$

30. (5), The distance between the two points is $\sqrt{(-4 - 5)^2 + (2 - 2)^2} = \sqrt{81} = 9$.

UNIT 4 GEOMETRY

LESSON 1, pp. 86–87

1. (1), Angles 1 and 5 are corresponding, and angles 5 and 6 are supplementary; $180° - 115° = 65°$

2. (1), The angles are supplementary; $180° - 100° = 80°$

3. (3), Vertical angles are congruent; $100°$

4. (2), $90° - 25° = 65°$

5. (4), Angles 1 and 3 are vertical angles, so they are congruent.

6. (3), Angles 1 and 3 are congruent because they are vertical. Angles 3 and 7 are congruent because they are corresponding.

7. (3), $\angle RMP$ and $\angle PMN$ are supplementary; $180° - 90° = 90°$

LESSON 2, pp. 88–89

1. (3), $m\angle B = 55°$, $360° - 2(55°) = 250°$, $250° \div 2 = 125°$

2. (1), The base angles each have a measure of $180° - 115° = 65°$, so the missing angle of the triangle is $180° - 65° - 65° = 50°$.

3. (4), Two angles are congruent, so two sides are congruent.

4. (3), Solve $x + 2x + 90° = 180°$ to find that $x = 30°$.

5. (3), $m\angle SUV = 180° - 45° = 35°$, and $\angle R$ is congruent to $\angle SUV$.

6. (4), $m\angle UST = 180° - 45° - 90° = 45°$, and $m\angle RSU = 180° - 135° = 45°$; $m\angle RSU + m\angle UST = 90°$

7. (5), Opposite angles are congruent; $360° - 35° - 35° = 290°$, and $290° \div 2 = 145°$.

LESSON 3, pp. 90–91

1. (3), $m\overline{AC} = 2 \times m\overline{RT}$; $2 \times 1.2 = 2.4$ meters

2. (4), $\triangle XYZ$ appears to have the same side lengths and angle measures as $\triangle EFG$.

3. (1), If the perimeter of triangle 1 is 19 ft, then $m\overline{EF}$ is $19 - 6 - 9 = 4$ ft, and $\overline{EF} \cong \overline{XY}$.

4. (3), Solve $\dfrac{2}{5.4} = \dfrac{4}{x}$ to find that $x = 10.8$ m.

5. (4), Let $x = m\overline{HG}$. Solve $\dfrac{2}{5} = \dfrac{4}{x}$ to find that $x = 10$ m. The perimeter is $4 + 10 + 10.8 = 24.8$ m.

6. (2), M corresponds to X, and N corresponds to Y, so $\overline{MN} \cong \overline{XY}$.

LESSON 4, pp. 92–93

1. (4), Solve $\dfrac{1 \text{ in.}}{2.5 \text{ mi}} = \dfrac{5 \text{ in.}}{x}$ to find that $x = 12.5$ miles.

2. (2), Use $\dfrac{\text{shadow}}{\text{tree}} = \dfrac{\text{shadow}}{\text{person}}$, or $\dfrac{1}{x} = \dfrac{2.5}{14}$.

3. (5), Since she drove there and back, she drove $2 \times 2.5 = 5$ cm. Solve $\dfrac{1 \text{ cm}}{6 \text{ km}} = \dfrac{5 \text{ cm}}{x}$ to find that $x = 30$ km.

4. (4), Solve $\dfrac{2 \text{ in.}}{4.8 \text{ mi}} = \dfrac{5.5 \text{ in.}}{x}$ to find that $x = 13.2$ miles.

5. (4), Solve $\dfrac{6}{8.5} = \dfrac{4.25}{x}$ to find that $x \approx 6.02$ feet.

6. (5), Solve $\dfrac{1 \text{ cm}}{20 \text{ km}} = \dfrac{1.5 \text{ cm}}{x}$ to find that $x = 30$ km.

7. (2), Jack: $\dfrac{1 \text{ cm}}{20 \text{ km}} = \dfrac{2.5 \text{ cm}}{x}$; $x = 50$ km. Pedro: $\dfrac{1 \text{ cm}}{20 \text{ km}} = \dfrac{2 \text{ cm}}{x}$; $x = 40$ km. Jack drove 10 km more.

8. (1), Carl drives a map distance of $2 + 2.25 + 2.25 + 2 = 8.5$ cm in all. Solve $\dfrac{1 \text{ cm}}{20 \text{ km}} = \dfrac{8.5 \text{ cm}}{x}$ to find that $x = 170$ km.

9. (3), Solve $\dfrac{12 \text{ in.}}{60 \text{ in.}} = \dfrac{1 \text{ in.}}{x}$ to find that $x = 5$ in., so 1 in. on the model represents 5 in.

LESSON 5, pp. 94–95

1. (5), Use $A = 3.14 \times r^2$ to find that $A = 3.14 \times 7^2 = 153.86$ cm².

2. (3), If $d = 15$, then $r = 7.5$ cm; $A = 3.14 \times 7.5^2 = 176.625$ cm².

3. (2), Use $C = 3.14 \times d$ to find that $C = 3.14 \times 25 = 78.5$ in.

4. (3), If $d = 18$, then $r = 9$ ft, and the area is $3.14 \times 9^2 = 254.34$ sq ft. The pavers will charge $\$1.59 \times 254.34 \approx \$404.40$.

5. (2), If $r = 7$, then $d = 14$ ft; use $C = 3.14 \times d$ to find that $C = 3.14 \times 14 = 43.96$ ft, which rounds to 44 ft.

6. (4), Use $A = 3.14 \times r^2$ to find that $A = 3.14 \times 4^2 = 50.24$ sq ft.

7. (5), Use $A = 3.14 \times r^2$ to find that $A = 3.14 \times 7^2 = 153.86$ sq ft, which rounds to 154 sq ft.

8. (3), First circle: $A = 3.14 \times 5.5^2 = 94.985$ sq cm. Second circle: $A = 3.14 \times 6.25^2 \approx 122.66$ sq cm. The second circle is about $122.66 - 94.985 = 27.675$ sq cm larger.

LESSON 6, pp. 96–97

1. (4), Use $V = 3.14 \times r^2 \times h$ to find that $V = 3.14 \times 9 \times 8 = 226.08$ in³.

2. (1), Use $V = \dfrac{1}{3} \times 3.14 \times r^2 \times h$ to find that $V = \dfrac{1}{3} \times 3.14 \times 4^2 \times 12$.

3. (5), Use $V = l \times w \times h$ to find that $V = 40 \times 20 \times 20 = 16{,}000$ in³; $16{,}000$ in³ $\times$ 50 bales $= 800{,}000$ in³.

4. (2), The first figure is a square pyramid, and the second figure is a cone.

5. (3), Use $V = 3.14 \times r^2 \times h$. Solve $9{,}156.24 = 3.14 \times 81 \times h$ for h; $h = 36$ in.

LESSON 7, pp. 98–99

1. (4), Add the areas of the two triangles and the square. Each triangle has an area of $\dfrac{1}{2} \times 3 \times 8 = 12$. The square has an area of $8 \times 8 = 64$; $12 + 12 + 64 = 88$ sq ft

2. (5), The two semicircles form one whole circle. Area of the circle $= 3.14 \times 2.5^2 = 19.625$; area of rectangle $= 8 \times 5 = 40$; $19.625 + 40 = 59.625$ sq ft

3. (3), The volume of one cone $= \dfrac{1}{3} \times 3.14 \times 2^2 \times 3 = 12.56$; $12.56 \times 2 = 25.12$ m³

4. (3), The volume of the cylinder $= 3.14 \times 2^2 \times 6 = 75.36$; $75.36 + 25.12 = 100.48$, which rounds to 100.

5. (5), The top rectangle is $16 \times 5 = 80$ sq ft, the middle rectangle is $10 \times 5 = 50$ sq ft, and the bottom rectangle is $5 \times 6 = 30$ sq ft; $80 + 50 + 30 = 160$ sq ft

6. (2), The perimeter is $16 + 5 + 6 + 11 + 5 + 6 + 5 + 10 = 64$ ft.

7. (5), Multiply the area by 3; $160 \times 3 = 480$ ft³

UNIT 4 (continued)

LESSON 8, pp. 100–101

1. (4), Solve $a^2 + 5^2 = 10^2$ for a to find that $a^2 = 75$, so $a \approx 8.66$ ft, which rounds to 8.7 ft.

2. (4), Solve $15^2 + 30^2 = c^2$ for c to find that $c^2 = 1,125$, so $c \approx 33.54$ ft.

3. (2), Solve $a^2 + 30^2 = 35^2$ for a to find that $a^2 = 325$, so $a \approx 18.03$ ft, which rounds to 18 ft.

4. (4), Solve $40^2 + 120^2 = c^2$ for c to find that $c^2 = 16,000$, so $c \approx 126.49$ m, which rounds to 126 m.

5. (1), Solve $20^2 + 120^2 = c^2$ for c to find that $c^2 = 14,800$, so $c \approx 121.655$ m, which rounds to 122 m.

6. (5), Plot the points on a grid; the distance between the points represents the hypotenuse; $d = \sqrt{(-4 - 4)^2 + (5 - 3)^2} = \sqrt{64 + 4} = \sqrt{68} \approx 8.25$, which rounds to 8.2.

UNIT 4 REVIEW, pp. 102–107

1. (5), m$\angle DBE$ + m$\angle EBC$ = 90°, so the angles are complementary.

2. (1), m$\angle ABE$ = 150° because m$\angle ABD$ = 90°, and m$\angle DBE$ = 60°, so 90° + 60° = 150°.

3. (1), Solve $\frac{1 \text{ cm}}{15 \text{ km}} = \frac{8.5 \text{ cm}}{x}$ to find that $x = 127.5$ km.

4. (3), The 120° angle and x are supplementary angles; 180° − 120° = 60°

5. (4), The 120° angle and y are vertical angles, so y = 120°.

6. (2), Use the formula $V = 3.14 \times r^2 \times h$. Solve $3,740 = 3.14 \times r^2 \times 17.5$ for r to find that $r^2 \approx 68.062$, so $r \approx 8.25$ cm. If $r \approx 8.25$ cm, then $d \approx 16.5$ cm.

7. (1), Angles 2 and 4 are vertical, so they have the same measure.

8. (4), Solve $a^2 + b^2 = c^2$ for c to find that $c^2 = 16,900$, so $c = 130$ yd. Wanda walked 130 yd instead of 50 + 120 = 170 yd, so she walked 170 − 130 = 40 fewer yd.

9. (2), The two half-circles form one whole circle. $P = 3.14(4.5) + 5 + 5 = 24.13$ ft.

10. (3), The circle is $A = 3.14 \times 2.25^2 \approx 15.896$ sq ft, and the rectangle is $A = 5 \times 4.5 = 22.5$ sq ft. The total area is 15.896 + 22.5 = 38.396 sq ft, which rounds to 38.40 sq ft.

11. (2), The sum of the interior angle measures of a triangle is 180°, so m$\angle G$ = 180° − 95° − 47° = 38°.

12. (4), Triangle HFG has three different-sized angles, and one angle is greater than 90°, so the triangle is obtuse.

13. (3), If d = 9 ft, then r = 4.5 ft, and $A = 3.14 \times 4.5^2 = 63.585$ sq ft, which rounds to 63.59 sq ft.

14. (4), If d = 9 ft, then $d = 9 \times 12$ in. = 108 in. The circumference is $3.14 \times 108 = 339.12$ in.

15. (3), Solve $\frac{1 \text{ in.}}{3 \text{ ft}} = \frac{8 \text{ in.}}{x}$ to find that the length is 24 ft, and solve $\frac{1 \text{ in.}}{3 \text{ ft}} = \frac{7 \text{ in.}}{x}$ to find that the width is 21 ft.

16. (2), Since $\angle C \cong \angle E$ because opposite angles are congruent, 360° − 125° − 125° = 110°, and 110° ÷ 2 = 55°.

17. (3), The sum of the interior angle measures of a triangle is 180°, so the measure of the missing angle is 180° − 90° − 37° = 53°.

18. (1), The volume of one cone is $V = \frac{1}{3} \times 3.14 \times 1.7^2 \times 4 \approx 12.09$ cubic inches. 480 cubic inches in all ÷ 12.09 cubic inches is 39.7, or about 40.

19. (1), Subtract the area of the grass from the area of the entire circle: $3.14 \times 11^2 = 379.94$; 379.94 − 253.3 = 126.64 sq ft

20. (100 sq ft), The diameter of the whole area is 22 + 5 + 5 = 32 ft. The circumference is $C = 3.14 \times 32 = 100.48$ ft, which rounds to 100.5.

21. (11.2), Multiply 5.6 miles by 2 to get a product of 11.2 miles.

22. (3), The volume of one pyramid is $\frac{1}{3} \times 12^2 \times 15 = 720$ cm^3, so the volume of two pyramids is $720 \times 2 = 1,440$ cm^3.

23. (4), Use the Pythagorean theorem; m$\overline{LN}$ = 3 units and m$\overline{NM}$ = 5 units; $3^2 + 5^2 = c^2$, so $c^2 = 34$ and $c \approx 6$.

24. (4), Solve $\frac{2 \text{ cm}}{29.8 \text{ km}} = \frac{2.8 \text{ cm}}{x}$ to find that $x = 41.72$ km.

25. (234.5), Lisa drove a map distance of 1.8 + 2.1 + 2.8 = 6.7 cm. Solve $\frac{1 \text{ cm}}{35 \text{ km}} = \frac{6.7 \text{ cm}}{x}$ to find that $x = 234.5$ km.

Index

Note: Page numbers in **boldface** indicate definitions or main discussion.

A

Acute angles, 88
Acute triangles, 88
Addition, **4**
  with decimals, 16
  with exponents, 64
  of fractions, 10
  of integers, 58–59
  inverse operation of, 62
Adjacent angles, **86**
Algebra, 57
  algebraic expressions, **60**, 60–61, 70
  coordinate grid, **72**, 72–73
  equations, **62**, 62–63
  exponents, **64**, 64–65
  factoring, 68–69
  functions, **66**, 66–67
  graphing equations, **74**, 74–75
  inequalities, **70**, 70–71
  integers, **58**, 58–59
  patterns, **66**, 66–67
  percentage of questions on, 57
  slope of a line, **76**, 76–77
  solving expressions, 60
  square roots, **64**
  variables, **60**, 60–61
Algebraic expressions, **60**, 60–61
  in inequalities, 70
  simplifying, 60
Alternate exterior angles, **86**
Alternate interior angles, **86**
Alternate-response format, 22, 50, 78, 102
Angles, **86**, 86–87
  acute/obtuse, 88
  adjacent, **86**
  alternate interior/alternate exterior, **86**
  classifying triangles by, 88
  complementary/supplementary, **86**
  congruent angles, **86**
  of congruent/similar figures, 90–91
  corresponding, 86
  exterior/interior, **86**
  naming, 86
  of parallelograms/rhombus, 88
  of quadrilaterals, 88
  right, 88
  of right triangles, 100
  of squares/rectangles, 88
  straight, 86
  of triangles, 88, 100
  vertex of, **86**
  vertical, **86**
Answers, checking, 62
Area, **34**, 34–35
  of circles, 94
  of irregular figures, 98
  of squares, 34
  units of measure for, 34, 94

B

Bar graphs, **44**, 44–45
Base of a number
  base-ten place value system, 14
  exponents and, **64**
  of percent problem, **20**
  in Scientific notation, 64
Bases of figures, 34, **34**
  of a cone, 96
  of a cylinder, 96
  of square pyramids, 96
  of trapezoids, 34
Base-ten place value system, 14

C

Calculator
  exponents with, 64
  keys on Casio FX-260 Solar, x, xi
  operations on Casio FX-260 Solar, xi
  square roots with, 64
  for use in GED test, vii, x
Capacity, 30. *See also* Volume
Centigrams, 30
Centiliter, 30
Centimeter, 30
Central measures
  mean, **38**, 38–39
  median, **38**, 38–39
  mode, **38**, 38–39
Change over time, 44
Change to figures, 72
Checking answers, 62
Circle graphs, **46**, 46–47
Circles, **94**, 94–95
  area of, 94
  circumference of, **32**, **94**
  radius/diameter of, **32**, **94**
Circumference, **32**, 32–33, **94**
Classification
  of quadrilaterals, 88
  of triangles, 88
Coefficients, 62
Combining like terms, 62
Commas in whole numbers, 2
Common denominators, 10
Commutative property, 36
Comparing
  decimals, 14
  whole numbers, 2
Complementary angles, **86**
Cones, 96
Congruent angles, 86
Congruent figures, 90–91, **91**
Conversion
  within metric system, 30
  of percents/fraction/decimals, 18–19

between systems of measure, 30
  within U. S. customary system, 30
Coordinate geometry
  coordinate grid, **72**, 72–73
  distance between two points, 100
  graphing a line, 74
  graphing equations, 74–75
  graphing inequalities, 70–71
  ordered pairs, **72**
  slope-intercept form of a line, 76
  slope of a line, 76
Coordinate grid, **72**, 72–73
Coordinate-grid-response format, 72, 78
Corresponding angles, 86
  of similar figures, 90
Cross products, 12
Cubes, **36**
  faces of, 96
  volume of, 36, 96–97
Cubic centimeters, 36
Cubic units, 36
Cup, 30
Customary system, 30–31
Cylinders, 96
  volume of, 96

D

Data analysis
  bar/line graphs, **44**, 44–45
  central measures, 38–39
  circle graphs, **46**, 46–47
  mean, **38**
  median, **38**
  mode, **38**
  probability, **40**, 40–41
  range, **38**
  tables, **42**, 42–43
Decimal point, 14, 16
  in percents, 18
Decimals, **14**, 14–15
  comparing/ordering, 14
  converting to/from fractions/percents, 18–19
  mixed numbers written as, 18
  multiplying by ten, 14
  operations with, 16–17
  probability expressed as, 40
  rounding, 14
  values on circle graphs as, 46
  written as fractions, 14
Denominator, **8**
  of slope of a line, 76
Dependent events, 40
Diameter, 32, **94**
Difference, **4**
Dilation, 72
Distance between two points
  finding with Pythagorean theorem, 100
  formula for, 74
Distance measurements, 32
Distance traveled, 48
Distributive property, **60**

L

Labels on graphs, 44, 46
Legs of right triangles, 100
Length, **32**, 32–33
    perimeter and, 32
    units of measure, 30
Less than or equal to symbol (≤), 70
Less than symbol (<), 2, 70
Like fractions, 10
Like signs, 58
Like terms, 60, 62
    operations with, 64
Linear equations, **74**
    slope of a line, 76
Line graphs, **44**, 44–45
    title/scale/axes, 44
Lines, **86**, 86–87
    drawing, 74
    linear equations generating, 74
    naming of, 86
    slope-intercept form of, **76**, 76–77
    slope of, **76**, 76–77
    symbol for (–), 86
Line segment, 72
Liter, 30

M

Mass, 30
Mathematical patterns, **66**, 66–67
Mean, **38**, 38–39
Measurement
    area, 34–35
    circumference, **32**, 32–33
    indirect, 90, **92**, 92–93
    length, **32**, 32–33
    perimeter, **32**, 32–33
    time, **48**, 48–49
    units/systems of
        measure, 30–31
    volume, **36**, 36–37
Measurement systems, **30**, 30–31
Median, **38**, 38–39
Meter, 30
Metric system, **30**, 30–31
Mile, 30
Milligrams, 30
Milliliter, 30
Millimeter, 30
Millions place, 2
Mixed numbers
    changing to/from improper
        fractions, 8
    converting to decimal/percent, 18
    operations with, 10
    ratios as, 12
Mode, **38**, 38–39
Multiple-choice questions, vii
Multiplication, **4**
    commutative property of, 36
    with decimals, 16

    with exponents, 64
    of factors, **68**
    of fractions, 10
    of integers, 58–59
    inverse operation of, 62
    by ten, 14
    ways of writing, 60

N

Naming
    angles, 86
    corresponding/similar figures, 90
    lines, 86
    triangles, 88
Negative exponents, 64
Negative numbers, 58–59
    in ordered pairs, 72
Negative sign (–), 58–59
Number line, 58
Number patterns, 66–67
Number sense, 4–5
    decimals, 14–15
    fractions, 8–9
    fractions/decimals/percents
        relationships, 18–19
    operations with decimals, 16–17
    operations with fractions, 10–11
    operations with whole
        numbers, 4–5
    percent problems, 20–21
    ratios/proportions, 12–13
    whole numbers, 2–3
    word problems, 6–7
Numerator, 8
    of slope of a line, 76

O

Obtuse triangles, 88
One
    as any number to zero power, 64
    as a coefficient, 62
    as denominator of unit rate, 12
    as denominator of whole
        numbers, 8
    as a power of a number, 64
Ones place, 2
Operations
    with decimals, 16–17
    with exponents, 64
    with fractions, 10–11
    with integers, 58–59
    inverse of, 62
    order of, 62
    with whole numbers, **4**, 4–5
Opposites, 58–59
Ordered pairs, **72**
    graphing a line with, 74
Ordering
    decimals, 14
    whole numbers, 2, 38

Order of operations, 60, 62
Order of terms, 60
Origin, **72**
Ounce, 30
Outcomes, 40
Output, 66

P

Parallelograms, **34**
    area of, 34
    sides/angles of, 88
Parentheses, 60
    changing expressions with, 74
Part (percentage of), 20
Patterns, **66**, 66–67
Per, 12
Percents, **18**
    converting mixed numbers to, 18
    converting to/from fractions/decimals,
        18–19
    formula for, 20
    of increase/decrease, 20
    probability expressed as, 40
    problems with, 18–19
    values on circle graphs as, 46
Percent sign (%), **18**
Perimeter, **32**, 32–33
    of irregular figures, 98
Pi (π), 32, 94
Pint, 30
Place value, 2
    of decimals, 14
Place value chart, 2, 14
Points, 72
    using two to find slope, 76
Polygons
    area, 34–35
    congruent/similar, **90**, 90–91
    perimeter of, **32**, 32–33
    See also Parallelograms; Quadrilaterals;
        Rectangles; Squares; Trapezoids;
        Triangles
Positive exponents, 64
Positive numbers, 58–59
Positive sign (+), 58
Pound, 30
Powers of numbers, 64
Probability, **40**, 40–41
    experimental, **40**
    expressed as a ratio/fraction/percent/
        decimal, 40
    theoretical, **40**
Product, 4, **68**
Proportions, **12**, 12–13
    cross products of, 12
    finding length of missing side of similar
        figures, 90
    indirect measurement using,
        92, 92–93
    solving percent problems with, 20
Pyramids, 96
Pythagorean theorem, **100**, 100–101

cylinders, 96
irregular, 98–99
pyramids, 96
rectangular prisms, 36
rectangular solids, 96
square pyramids, 96
volume of, **36**, 36–37, 96, 98.
    See also Capacity
Time, **elapsed, 48**
Title
of graphs, 44
of tables, 42
Ton, 30
Translations, 72
Trapezoids, **34**
area of, 34
sides of, 88
Triangles, 86
acute, 88
area of, 34
classification of, 88
equilateral, 88
isosceles, 88
naming, 88
obtuse, 88
perimeter of, 32
Pythagorean theorem and, **100**,
    100–101
right, 88, **100**, 100–101
scalene, 88
similar, 90
sum of angles of, 88
Two-dimensional figures
area of, **34**, 34–35, 94, 98
bases of, 34
circles, **32**, 32–33, **94**, 94–95
congruent/similar, **90**, 90–91
indirect measurement of, **92**, 92–93,
    100, 100–101
irregular, **98**, 98–99
parallelograms, 34, 88
perimeter of, **32**, 32–33, 98
quadrilaterals, **88**, 88–89
rectangles, 32, 34, 88
squares, 34, 88
trapezoids, 34, 88
triangles, 32, 34, **88**, 88–89, 90, **100**,
    100–101

U

Unit rates, 12
Unit reviews, 22–27, 50–55, 78–83,
    102–107
Units of measure, **30**, 30–31
for area, 34
for volume, 36
Unlike fractions, 10
Unlike signs, 58
Unlike terms, 60
U. S. customary system, 30

V

Variables, **60**, 60–61
in equations, 62
grouping while solving equations, 62
two, 74
Vertex
of an angle, **86**
of a cone, 96
of square pyramids, 96
Vertical angles, **86**
Vertical axis, 44
Vertical line
in coordinate grid, 72
slope of, 76
Volume, **36**, 36–37
of irregular figures, 98
units of measure for, 30, 36

W

Weight, 30
Whole numbers, **2**, 2–3
comparing/ordering, 2
decimals and, 14
integers, 58–59
ordering, 38
rounding, 2
values on circle graphs as, 46
writing as fractions, 10
Word problems, **6**, 6–7
multiple-step solutions, 34
Writing
proportions, 12
ratios as fractions, 12
whole numbers, 2
whole numbers as fractions, 10

X

X-axis, 72
X-value, 66, 72

Y

Yard, 30
Y-axis, 72
Y-intercept, 76
Y-value, 66, 72

Z

Zero
as a placeholder, 4, 14
as a power of a number, 64
as slope of line, 76